SpringerBriefs in Computer Science

SpringerBriefs present concise summaries of cutting-edge research and practical applications across a wide spectrum of fields. Featuring compact volumes of 50 to 125 pages, the series covers a range of content from professional to academic.

Typical topics might include:

- A timely report of state-of-the art analytical techniques
- A bridge between new research results, as published in journal articles, and a contextual literature review
- A snapshot of a hot or emerging topic
- An in-depth case study or clinical example
- A presentation of core concepts that students must understand in order to make independent contributions

Briefs allow authors to present their ideas and readers to absorb them with minimal time investment. Briefs will be published as part of Springer's eBook collection, with millions of users worldwide. In addition, Briefs will be available for individual print and electronic purchase. Briefs are characterized by fast, global electronic dissemination, standard publishing contracts, easy-to-use manuscript preparation and formatting guidelines, and expedited production schedules. We aim for publication 8–12 weeks after acceptance. Both solicited and unsolicited manuscripts are considered for publication in this series.

**Indexing: This series is indexed in Scopus, Ei-Compendex, and zbMATH **

Long Zhao • Hongrui Shen • Kan Zheng

AI for Wireless Physical Layer

 Springer

Long Zhao
School of Information and Communication
Engineering
Beijing University of Posts
&Telecommunications
Beijing, China

Hongrui Shen
School of Information and Communication
Engineering
Beijing University of Posts
&Telecommunications
Beijing, China

Kan Zheng
College of Electrical Engineering
and Computer Sciences
Ningbo University
Zhejiang, China

ISSN 2191-5768   ISSN 2191-5776   (electronic)
SpringerBriefs in Computer Science
ISBN 978-3-032-01366-8   ISBN 978-3-032-01367-5   (eBook)
https://doi.org/10.1007/978-3-032-01367-5

This Springer imprint is published by the registered company Springer Nature Switzerland AG
The registered company address is: Gewerbestrasse 11, 6330 Cham, Switzerland

If disposing of this product, please recycle the paper.

Preface

The rapid evolution of Artificial Intelligence (AI) has catalyzed transformative breakthroughs in wireless network development. AI-empowered network architectures have emerged as a strategically critical research frontier, particularly in addressing the complex technical demands of Sixth-Generation (6G) mobile communication systems and infrastructures. Contemporary research initiatives primarily focus on three dimensions: intelligent optimization of legacy algorithms through machine learning paradigms, cross-layer protocol stack enhancements enabled by neural networks, and cognitive network management systems employing deep reinforcement learning. These advancements leverage AI's unparalleled competencies in multidimensional pattern recognition and adaptive decision-making, enabling 6G networks to achieve comprehensive performance improvements spanning physical-layer signal processing, network-layer resource allocation, and service-layer quality-of-experience optimization.

The objective of this monograph is to present the paradigm of AI for wireless physical layer and to examine several typical 6G applications by leveraging AI's capabilities in feature extraction and logical deduction. In Chap. 1, we present the significant meaning of AI-based networks or future communications after a brief introduction to the scenarios and capabilities of future 6G networks. Chapter 2 develops a two-stage deep learning framework for channel estimation: intelligent pilot-position channel estimation followed by time-frequency interpolation. On the one hand, the estimated Channel State Information (CSI) is obtained at the User Equipment (UE) under Frequency Division Duplex (FDD) systems, therefore intelligent CSI compression and feedback is correspondingly studied in order to reduce the feedback overhead for 6G system in Chap. 3. On the other hand, based on the estimated channel at The next Generation Node B (gNB) under Time Division Duplex (TDD) systems, the downlink precoder could be intelligently generated by designed deep learning network in Chap. 4. Moreover, the beamforming-enabled mm-wave technology should be adopted to enhance system transmission rate, therefore Chap. 5 investigates the deep learning-based spatial beam management in order to minimize the measurement overhead of beam pairs. Finally, Chap. 6

suggests some open issues that help point out some new research directions in AI-based communications.

We are very grateful to Prof. Xuemin (Sherman) Shen, the *Springer Briefs* series editor on Wireless Communications. This book would not be possible without his kind support. Special thanks are also attributed to Susan Lagerstrom-Fife and Sudha Ramachandran at Springer Science+Business Media, for their assistance throughout the preparation process of this monograph.

We would like to thank Yan Zhao, Jiayi Xu, Xun Zhang, Jilong He, Chunhuan Wang, Jiawei Li, Zhan Peng, Hanqing Xu, Zibo He, and Xinfang Chen from the Wireless Signal Processing and Network (WSPN) group at the Beijing University of Posts and Telecommunications (BUPT) for their contributions to this monograph. We also would like to thank all the members of the WSPN group for their thought-provoking discussions and insightful suggestions, creative ideas, and valuable comments.

This work is funded in part by the National Science Foundation of China.

Beijing, China   Long Zhao
Beijing, China   Hongrui Shen
China   Kan Zheng

Declarations

Competing Interests The authors have no competing interests to declare that are relevant to the content of this manuscript.

Contents

Acronyms

2D	Two-Dimensional
3D	Three-Dimensional
3GPP	3rd Generation Partnership Project
5G	Fifth-Generation
5G-NR	Fifth-Generation New Radio
6G	Sixth-Generation
Adam	Adaptive Momentum
AI	Artificial Intelligence
AI4NET	AI for Network
AoA	Azimuth of Arrival
AoD	Azimuth of Departure
AR	Augmented Reality
AWGN	Additive White Gaussian Noise
BD	Block Diagonalization
BiLSTM	Bidirectional Long Short Term Memory
BLER	Block Error Rate
BN	Batch Normalization
BNA	Bit Number Adjusting
CAM	Channel Attention Module
CDF	Cumulative Distribution Function
CDL	Clustered Delay Line
CNN	Convolutional Neural Network
CP	Cyclic Prefix
CPU	Central Processing Unit
CR	Compression Ratio
CR	Correlation Rotation
CRM	Channel Reconstruction Module
CS	Compressive Sensing
CSI	Channel State Information
CV	Computer Vision
DnCNN	Denoising Convolutional Neural Network

DL	Deep Learning
DRX	Discontinuous Reception
DT	Decision Tree
EDBT	Enhanced DNN-based Beam Training
ELU	Exponential Linear Unit
eMBB	Enhanced Mobile Broadband
eTypeII	Enhanced Type II
EVD	Eigenvalue Decomposition
FBCU	Feedback Bit Control Unit
FC	Fully Connected
FCN	Fully Connected Network
FDD	Frequency Division Duplex
FFEM	Fine Feature Extraction Module
FLOPs	Floating Point Operations
gNB	The next Generation Node B
GNB	Gaussian Naive Bayes
GNNs	Graph Neural Networks
GPU	Graphics Processing Unit
IMT	International Mobile Telecommunications
INet	Interpolation Network
IoT	Internet of Things
ITU-R	International Telecommunication Union Radiocommunication Sector
LDPC	Low Density Parity Check
LMMSE	Linear Minimum Mean Square Error
LN	Layer Normalization
LOS	Line of Sight
LR	Logistic Regression
LS	Least Square
LSTM	Long Short Term Memory
MAC	Media Access Control
MET	Multiuser Eigenmode Transmission
MIMO	Multiple-Input and Multiple-Output
ML	Machine Learning
MM	Mixer Module
MMSE	Minimum Mean Square Error
mMTC	Massive Machine Type Communications
MLP	Multi-Layer Perceptron
MSE	Mean Square Error
MU-MIMO	Massive Multiuser Multiple-Input and Multiple-Output
NET4AI	Network for AI
NLOS	Non-Line of Sight
NMSE	Normalized Mean Square Error
ODBT	Original DNN-based Beam Training
OFDM	Orthogonal Frequency Division Multiplexing

OM	Operational and Maintenance
PCENet	Pilot-Position Channel Estimation Network
PDSCH	Physical Downlink Shared Channel
PHY	Physical Layer
PM	Pre-patch Module
PUM	Previous Up-sampling Module
QoS	Quality of Service
QPSK	Quadrature Phase Shift Keying
RB	Resource Block
ReLU	Rectified Linear Unit
RF	Random Forest
RFEM	Rough Feature Extraction Module
RL	Reinforcement Learning
RRC	Radio Resource Control
RSRP	Reference Signal Received Power
RSRQ	Reference Signal Receiving Quality
SGCS	Square of Generalized Cosine Similarity
SGD	Stochastic Gradient Descent
SNR	Signal-to-Noise Ratio
SRCNN	Super Resolution Convolutional Neural Network
SVM	Support Vector Machine
TDD	Time Division Duplex
UAV	Unmanned Aerial Vehicle
UE	User Equipment
UGI	Ubiquitous General Intelligence
UPAs	Uniform Planar Arrays
UQ	Uniform Quantization
URLLC	Ultra-Reliable and Low Latency Communications
VQ	Vector Quantization
VR	Virtual Reality
WMMSE	Weighted Minimum Mean Square Error
WRC	World Radiocommunication Conference
ZF	Zero-Forcing

Chapter 1
Introduction

Abstract AI-based networks constitute a pivotal and timely research area, driven by the technical requirements of 6G and future wireless communication systems. This chapter first outlines the motivations for AI-enhanced 6G technologies, following a concise overview of 6G's performance requirements and capacity benchmarks. We then introduce core concepts and application paradigms for both AI for network (AI4NET) and network for AI (NET4AI), concluding with a clear articulation of the monograph's objectives.

Keywords AI · 6G · Physical layer · AI4NET · NET4AI

1.1 Sixth-Generation (6G) Brief

The advancement of communication technology is inseparable from the massive demand for connectivity in the information society and the expansion of related industrial ecosystems. Since the late 1980s, mobile communication systems have evolved through five generations of technological iterations, largely adhering to a decadal generational cycle. Each generation has been designed to satisfy the communication needs of the information society at their respective times and for the foreseeable future. Future Sixth-Generation (6G) mobile communication systems will continue this trajectory of development, treating networks and terminals as a unified entity. The intelligent capabilities of terminals will be further explored and realized, serving as the foundation for technological planning and evolutionary roadmaps across the physical layer, network layer, and service layer. This holistic approach will drive innovations in network-terminal synergy, enabling adaptive resource allocation, context-aware service provisioning, and embedded Artificial Intelligence (AI) functionalities to meet escalating demands for ultrareliable, low latency, and intelligent connectivity in emerging applications [1].

L. Zhao et al., *AI for Wireless Physical Layer*, SpringerBriefs in Computer Science,
https://doi.org/10.1007/978-3-032-01367-5_1

1.1.1   Scenarios of 6G Networks

Looking ahead to 2030 and beyond, the International Telecommunication Union Radiocommunication Sector (ITU-R) is committed to advancing International Mobile Telecommunications (IMT)-2030 (6G). The finalized overarching timeline for 6G development, as agreed by ITU-R, comprises three phases:

Phase 1: Completion of vision finalization by June 2023, ahead of the World Radiocommunication Conference (WRC)-23;

Phase 2: Definition of requirements and evaluation methodology by 2026;

Phase 3: Delivery of specifications by 2030.

In June 2023, a draft recommendation for IMT-2030 was unanimously adopted during a meeting held in Geneva, Switzerland. As illustrated in Fig. 1.1, this framework defines six key usage scenarios for IMT-2030. Building on the "iron triangle" of IMT-2020 (Fifth-Generation (5G))—Enhanced Mobile Broadband (eMBB), Ultra-Reliable and Low Latency Communications (URLLC), and Massive Machine Type Communications (mMTC)—IMT-2030 expands this concept into a hexagonal model. Encircling the hexagon, four cross-cutting design principles applicable to all scenarios are highlighted: sustainability, ubiquitous intelligence, security/privacy/resilience, and connecting the unconnected. These principles aim to guide the development of 6G technologies toward inclusive, intelligent, and secure global connectivity [2].

Fig. 1.1 Illustration of six key usage scenarios for IMT-2030

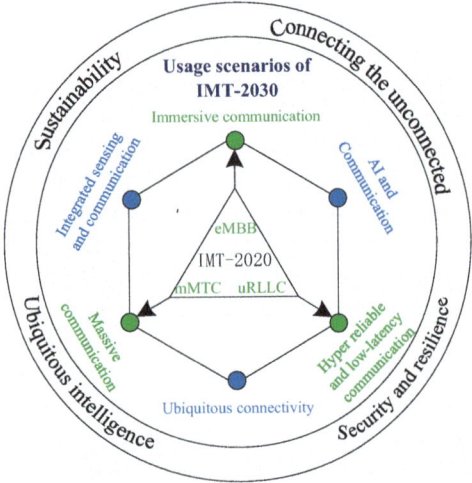

Fig. 1.2 Capabilities of
IMT-2030

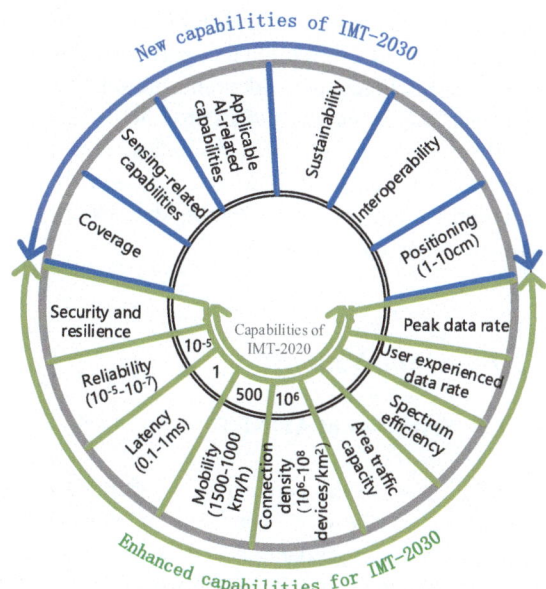

1.1.2 Capabilities of 6G Networks

From a communication rate perspective, 6G will advance from Gbit/s to Tbit/s. In terms of information scope, it will expand from terrestrial mobile communication to omnidirectional connectivity spanning land, sea, air, and space. Regarding network services, it will refine communication intelligence, transitioning from human-machine-thing interactions to integrated intelligent entities that seamlessly merge humans, machines, physical objects, and artificial intelligence.

As illustrated in Fig. 1.2, the capability requirements for IMT-2030 encompass nine enhanced capabilities: peak data rate, user-experienced data rate, spectrum efficiency, area traffic capacity, connection density, mobility, latency, reliability and security, and privacy and resilience. Additionally, it introduces six new capabilities: coverage, positioning accuracy, sensing-related functionalities, AI-native integration, sustainability, and interoperability. These metrics collectively aim to redefine global connectivity by balancing ultrahigh performance with adaptive, intelligent, and universally inclusive network architectures [2].

1.2 AI-Based 6G Technology

The evolution of AI technologies has profoundly accelerated the advancement of wireless networks. AI for Network (AI4NET) denotes the systematic application of AI to optimize wireless systems, aiming to elevate network performance, operational efficiency, and quality of service. This paradigm centers on AI-driven enhancements to conventional algorithms, functional upgrades across protocol layers, and intelligent network orchestration. Conversely, Network for AI (NET4AI) represents the reciprocal paradigm where 6G networks empower AI by providing distributed computational resources, federated data pipelines, and embedded security frameworks. NET4AI focuses on enabling real-time AI training/inference with minimal latency while ensuring data security and privacy preservation through network-native mechanisms [3, 4].

1.2.1   AI for Network

1.2.1.1   Overview of AI4NET

AI4NET refers to the application of artificial intelligence to enhance the performance and user service experience of network systems [5]. By leveraging AI's ability to analyze vast datasets and execute real-time decision-making, it addresses dynamic challenges in wireless networks, such as fluctuating environmental conditions and high-dimensional data variability. Key research areas include optimizing traditional algorithms, refining network functions, and improving operational and Operational and Maintenance (OM) management. These advancements aim to boost transmission efficiency and wireless performance across both air interface and network infrastructure. AI models thrive in this context by establishing robust input-output mappings from sparse, time-variant data, while continuous data generation ensures sufficient training samples for adaptive learning and proactive system adjustments.

1.2.1.2   Specific Applications of AI4NET

AI4NET demonstrates versatile applications. Under improved OM efficiency, AI enables traffic prediction algorithms for proactive maintenance and network parameter optimization through digital twin technology. In air interface performance enhancement, it supports critical functions like Channel State Information (CSI) feedback for feature extraction, adaptive Orthogonal Frequency Division Multiplexing (OFDM) receiver design, and beam management algorithms to conserve base station energy. Additionally, AI-driven mobility management and network slicing optimize resource allocation and service quality. For new business enablement, real-time wireless localization algorithms and environment-aware systems unlock innovative services. By dynamically allocating resources (e.g., bandwidth, power) and addressing data sparsity, AI ensures resilient and adaptive wireless network operations, paving the way for intelligent, self-optimizing systems [6].

1.2.2 Network for AI

1.2.2.1 Overview of NET4AI

NET4AI represents a revolutionary paradigm in the deep convergence of 6G and AI, fundamentally redefining network architecture to embed native AI capabilities at its core. As a pivotal component of Phase II in 6G-AI integration, it transcends traditional communication networks limited to connectivity services, evolving into a dynamic, multidimensional platform that seamlessly integrates connectivity, distributed computing power, federated data ecosystems, and adaptive AI algorithms [7]. By embedding intelligence endogenously across core networks, access networks, and edge devices, NET4AI orchestrates resources such as real-time data streams, distributed computational nodes, and pretrained model repositories to form an "intelligence fabric." This architecture enables on-demand resource allocation, context-aware task offloading, and AI-driven network optimization. Built with native support for privacy-preserving mechanisms (e.g., federated learning, homomorphic encryption) and energy-efficient AI workflows, NET4AI ensures secure, low-latency execution of distributed training and inference tasks. It lays the foundation for Ubiquitous General Intelligence (UGI) by transforming networks into self-adaptive systems capable of autonomous decision-making, continuous learning, and proactive service provisioning across heterogeneous environments [8].

1.2.2.2 Specific Applications of NET4AI

The transformative potential of Net for AI manifests across diverse domains: In autonomous systems, it enables real-time collaborative inference among vehicles and edge servers for split-second decision-making in complex traffic scenarios. For industrial Internet of Things (IoT), it supports cross-factory federated learning, allowing manufacturers to collaboratively refine AI models while maintaining data sovereignty. In healthcare, NET4AI facilitates secure, decentralized medical imaging analysis by integrating edge computing with differential privacy frameworks. The architecture also empowers immersive metaverse experiences through distributed rendering engines and AI-driven avatar interactions, dynamically balancing computational loads across cloud-edge-device tiers. Smart cities leverage its predictive resource orchestration to optimize energy grids and traffic flows using federated data from sensors and user devices. Additionally, NET4AI drives advancements in personalized AI services—such as context-aware digital twins and adaptive Augmented Reality (AR)/Virtual Reality (VR) interfaces—by intelligently synchronizing localized inference with centralized model updates. These applications underscore its role as a unified platform that harmonizes connectivity, computation, data, and algorithms, accelerating the transition from fragmented AI solutions to pervasive, service-oriented intelligence ecosystems [6, 8].

1.3   Aim of Monograph

While the foundational architectures and numerous application scenarios of AI4NET and NET4AI have been extensively explored in existing research, this monograph primarily concentrates on the technological aspects of AI4NET. Although preliminary evaluations of diverse AI4NET application scenarios have been conducted, yielding several empirical outcomes, the design methodologies and detailed architectures of the corresponding deep learning networks or models remain under-explored and lack systematic synthesis in the current literature.

This monograph presents the fundamental applications of AI in physical layer communications, along with the essential design methodologies and detailed network architectures for these implementations. Channel estimation is the prerequisite for beamforming or precoding and will be first investigated based on AI technology, where both estimation and interpolation networks are designed and evaluated, respectively. Furthermore, under Frequency Division Duplex (FDD) operation, where downlink channel estimation is performed at User Equipment (UE)s, the monograph examines AI-enhanced feedback mechanisms to efficiently relay CSI from UEs to The next Generation Node B (gNB)s, enabling downlink precoding generation. Subsequently, to generate downlink precoding, the monograph details AI-driven techniques for downlink precoding construction at gNBs, which are critical for optimizing link transmission performance. Finally, AI-based beam management is discussed under millimeter-wave scenario for UE accessing to the network, which is the key technology for future 6G networks.

References

1. Harun, U.R., Seong, H.J.: AI empowered 6G technologies and network layers: recent trends, opportunities, and challenges. Expert Syst. Appl. **267**, 125985 (2025)
2. ITU-R: Framework and overall objectives of the future development of IMT for 2030 and beyond, pp. 1–19. ITU Publications (2023)
3. Yue, L., Chen, T.: AI large model and 6G network. In: 2023 IEEE Globecom Workshops (GC Wkshps), pp. 2049–2054 (2023)
4. Pan, J., Cai, L., Yan, S., Shen, X.S.: Network for AI and AI for network: challenges and opportunities for learning-oriented networks. IEEE Netw. **35**(6), 270–277 (2021)
5. Chen, T., Deng, J., Tang, Q., Liu, G.: Optimization of quality of AI service in 6G native AI wireless networks. Electronics **12**(15), 3306 (2023)
6. Cui, Q., You, Y., Ni, W., et al.: Overview of AI and communication for 6G network: fundamentals, challenges, and future research opportunities, pp. 1–74 (2024). CoRR abs/2412.14538
7. Mahmoud, H., Elbadawy, H.M., Ismail, T., Mi, D.: A comprehensive review of generative AI applications in 6G. In: 6th Novel Intelligent and Leading Emerging Sciences Conference (NILES), pp. 593–596 (2024)
8. Tong, W., Li, G.Y.: Nine challenges in artificial intelligence and wireless communications for 6G. IEEE Wirel. Commun. **29**(4), 140–145 (2022)

Chapter 2
Intelligent Channel Estimation Technology

Abstract This chapter explores a two-stage intelligent channel estimation scheme for OFDM systems, encompassing pilot-position channel estimation and full temporal-frequency-domain channel interpolation. In the first stage, a CNN-based pilot-position channel estimation network (PCENet) is designed. It extracts and learns features from the transmitted and received signal matrices at pilot positions, yielding the CSI matrix for these positions. The second stage leverages the pilot-position estimation results to design a fully connected network (FCN)-based full temporal-frequency-domain interpolation network (INet). This network learns the correlation features between frequency-domain subcarriers and time-domain symbols, ultimately outputting the full temporal-frequency-domain CSI matrix.

Keywords AI · OFDM · Pilot · Channel estimation · Interpolation

2.1 Background

Benefiting from the generation of massive data and the continuous improvement of computational capabilities in machine equipment, AI has flourished in recent years, with applications visible across various industries. As a core branch of AI, Deep Learning (DL) has also witnessed a surge in development, achieving remarkable success in fields such as image classification, machine translation, and speech recognition, gradually replacing traditional algorithms. In wireless communications, the integration of DL has become a global research focus in academia and industry, primarily in areas including physical layer transmission [1], wireless resource scheduling [2], congestion control [3], and semantic communication [4]. Through recent research, the industry has drawn the following key conclusions: (1) DL can characterize and reconstruct wireless channel environments, fully exploiting statistical properties of channel spaces to enhance the performance of physical layer algorithms. (2) DL enables global resource optimization for multiuser and multi-objective systems by leveraging communication, sensing, and computational resources. (3) DL can integrate with wireless network topologies and transmission protocols to construct novel intelligent network architectures, better serving future communication systems.

Channel estimation, as a critical technology in the physical layer, has also garnered significant attention. The conventional channel estimation algorithms for pilot positions include Least Square (LS) estimation, Minimum Mean Square Error (MMSE) estimation [5], and Linear Minimum Mean Square Error (LMMSE) estimation [6]. In order to balance the estimation performance and complexity of the LS, MMSE, and LMMSE algorithms, many researchers have conducted intensive studies. For insufficient estimation accuracy of the LS algorithm, [7] consider the influence of frequency offset and add frequency compensation to the traditional LS algorithm. And [8] transform the LS estimation results to the wavelet domain and design a threshold denoising function to denoise the channel in the wavelet domain. [9] focus on local errors and reduce the overall error by compensating for each estimated channel coefficient, thereby reducing the bit error rate. In order to reduce the complexity of the MMSE and LMMSE algorithms, [10] utilize singular value decomposition to simplify the derivation formula of MMSE from a mathematical perspective. [11] start from the correlation matrix and simplify the calculation of the correlation matrix at the current moment by using the CSI of the previous moment and the time characteristics of the channel. For large-scale multiuser Multiple-Input and Multiple-Output (MIMO) systems, [12] utilize the low-rank characteristic of the channel covariance matrix and reduce the dimension of the covariance matrix by subspace projection of the two-step covariance matrix, thereby reducing the computational load.

With the development of AI, researchers have introduced AI into channel estimation, aiming to improve estimation accuracy or reduce computational complexity through deep learning. An Fully Connected Network (FCN) was designed for channel estimation and signal detection, replacing traditional OFDM receivers with neural networks to merge channel estimation and signal detection into a single step [13]. However, due to its purely data-driven nature without incorporating existing communication algorithms, this approach suffers from issues such as lack of interpretability, slow convergence, and high computational costs. To address these limitations, a model-driven neural network named ComNet was proposed, comprising two subnetworks: a channel estimation subnetwork using LS estimation results as input to generate full temporal-frequency-domain CSI matrices, and a signal detection subnetwork producing recovered bits from Zero-Forcing (ZF) equalization results of the temporal-frequency-domain estimation matrix [14]. This work demonstrates that deep learning can synergize with traditional communication algorithms to enhance performance while reducing complexity. For millimeter-wave (mmWave) massive MIMO systems, an iterative channel estimation method was developed, employing the image denoising network Denoising Convolutional Neural Network (DnCNN) to update CSI estimates during each iteration [15, 16]. To reduce computational overhead at User Equipment (UE) with weaker computing power, leveraging the low-rank characteristics of massive MIMO systems, a novel approach where UEs directly feed back received pilot signals to computationally powerful base stations for estimation with an FCN-based estimator was proposed in [17]. Focusing on the interpolation process, [18] treat the pilot-position CSI matrix estimated by the LS algorithm as a low-resolution noisy image and use it as network input. Leveraging the classical super-resolution network Super Resolution Convolutional Neural Network (SRCNN) [19] from the image processing field, the method upsamples the low-dimensional pilot-position CSI matrix. Subsequently, DnCNN [16] is applied to denoise the upsampled matrix, ultimately yielding the full time-frequency-domain CSI matrix. To address the issue of excessive parameters in FCNs, a Convolutional Neural Network (CNN) is developed for channel estimation in mmWave massive MIMO systems by exploiting spatial and frequency correlations of channels [20]. The CNN takes CSI matrices from multiple consecutive subcarriers as input and outputs full time-frequency-domain CSI after channel feature extraction. For vehicular mmWave systems, a two-stage network that initially employs an FCN for channel estimation is designed, followed by feeding the results into a Bidirectional Long Short Term Memory (BiLSTM) network for channel tracking [21]. Moreover, targeting vehicular networking scenarios, a neural network incorporating Long Short Term Memory (LSTM) and FCN to enhance estimation performance is designed in [22].

In OFDM systems, pilot-based channel estimation comprises two stages: pilot-position channel estimation and full temporal-frequency-domain channel interpolation. For the former, conventional algorithms often struggle to balance system performance and complexity, failing to effectively integrate temporal and frequency-domain internal correlation variations or completely eliminate noise impacts. For the latter, interpolation itself constitutes a function-fitting process that cannot substantially improve overall channel estimation accuracy. To contrast with traditional algorithms, this chapter proposes a two-stage deep channel estimation model, designing the Pilot-Position Channel Estimation Network (PCENet) and the full temporal-frequency-domain deep Interpolation Network (INet), respectively. As shown in Fig. 2.1, the first stage involves deep channel estimation at pilot positions, where a CNN-based network is designed to extract and learn features from pilot-position transceiver signals. This network primarily consists of three modules: the coarse feature extraction module, the fine feature extraction module, and the channel attention module, ultimately outputting the pilot-position CSI matrix. The second stage implements deep interpolation across the full temporal-frequency domain using an FCN-based method, which includes the pre-upsampling module, pre-patch module, Mixer module, and channel reconstruction module, finally generating the complete temporal-frequency-domain CSI matrix.

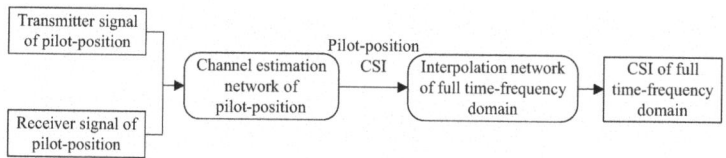

Fig. 2.1 Overall framework for two-stage full temporal-frequency channel estimation

2.2   CNN-Based Channel Estimation

2.2.1   Overall Network Architecture

This section analogizes the CSI matrix to an image, where each temporal-frequency grid corresponds to a pixel. Capitalizing on the unparalleled advantages of CNN in learning pixel features, this section designs the CNN-based PCENet to learn the mapping relationship between pilot-position transceiver signals and pilot-position CSI, thereby obtaining channel responses at pilot positions.

As shown in Fig. 2.2, PCENet employs CNN as the primary processing unit and incorporates three specifically designed modules for pilot-position channel estimation tasks. From input to output, these modules are sequentially arranged as: the Rough Feature Extraction Module (RFEM), Fine Feature Extraction Module (FFEM), and Channel Attention Module (CAM). Prior to outputting the pilot-position CSI matrix, a 1×1 convolutional kernel is applied to reduce the channel dimension of the CSI matrix. Furthermore, PCENet introduces residual connections between the RFEM and CAM, aiming to alleviate training challenges in deep networks while enhancing estimation performance.

2.2.2   Submodule Designs

The input of PCENet is the transceiver signal matrix at pilot positions, with dimensions $M_P \times N_P \times 2$, where M_P and N_P represent the number of subcarriers and OFDM symbols in the pilot pattern, respectively, and the value 2 corresponds to the transmitted and received signals. Since mainstream neural networks currently only process real numbers for computation and training, the real and imaginary parts of the complex-valued channel are concatenated along the channel dimension, ultimately forming an input matrix H_{TR} with dimensions $M_P \times N_P \times 4$.

Fig. 2.2 PCENet overall structure

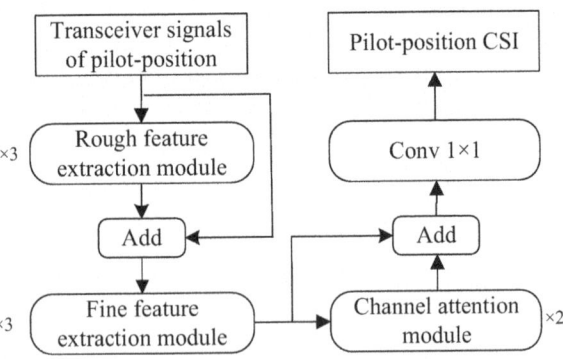

2.2.2.1 Rough Feature Extraction Module

The input matrix $\mathbf{H}_{TR}$ first undergoes RFEM to coarsely extract features of pilot-position transceiver signals, yielding a coarse feature matrix with dimensions $M_P \times N_P \times C$:

$$\mathbf{H}_{RF} = f_{RFEM}(\mathbf{H}_{TR}), \tag{2.1}$$

where C denotes the elevated channel dimension for enhanced extraction and learning of channel features, and $f_{RFEM}(\cdot)$ represents the rough feature extraction module.

Figure 2.3 illustrates the detailed network structure of the RFEM. After the pilot-position transceiver signal matrix $\mathbf{H}_{TR}$ is fed into this module, it first passes through a 1×3 convolutional layer to perform shallow extraction of correlations between OFDM symbols. Subsequently, sequential 5×1, 7×1, and 9×1 convolutional layers are applied to learn subcarrier-wise correlations. Finally, a lateral connection summation operation generates the coarse feature matrix $\mathbf{H}_{RF}$. The reason why the size of the convolution kernel used to extract OFDM symbol correlation is smaller than that used to extract subcarrier correlation is that under the selected pilot pattern, the number of OFDM symbols is much smaller than the number of subcarriers (i.e., $N_P \ll M_P$). Larger kernels for OFDM symbol processing would only introduce zero-padding artifacts at symbol boundaries without enhancing feature extraction, while unnecessarily increasing computational overhead.

Furthermore, the separation of OFDM symbol and subcarrier correlation extraction is motivated by the weak intrinsic coupling between frequency-domain subcarriers and time-domain symbols in OFDM systems. Mixed Two-Dimensional (2D) convolutional kernels (e.g., 3×3, 5×5) would create interference between temporal and spectral feature learning, thereby degrading estimation accuracy.

2.2.2.2 Fine Feature Extraction Module

Next, the coarse features $\mathbf{H}_{RF}$ undergo fine feature extraction to obtain a refined feature matrix with identical dimensions:

$$\mathbf{H}_{FF} = f_{FFEM}(\mathbf{H}_{RF}), \tag{2.2}$$

where $f_{FFEM}(\cdot)$ denotes the fine feature extraction module.

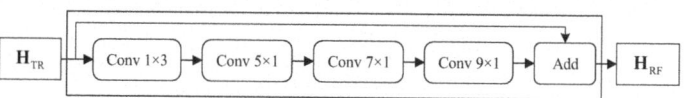

Fig. 2.3 RFEM structure

Figure 2.4 illustrates the detailed network structure of the fine feature extraction module. Unlike the rough extraction module, this module separates subcarrier correlation learning and OFDM symbol processing into two independent branches to further prevent mutual interference between these dimensions. The subcarrier branch consists of 5×1, 7×1, and 9×1 convolutional layers, while the OFDM symbol branch contains a 1×3 convolutional layer. After correlation learning and feature representation, outputs from both branches are concatenated along the channel dimension to form the final refined feature matrix $\mathbf{H}_{FF}$. The reason for adopting concatenation instead of additive fusion used in the rough extraction module is twofold: Concatenation increases the channel dimension of feature maps to enhance high-level channel feature learning, while simultaneously improving model expressiveness by diversifying network processing units, thereby avoiding redundant stacking of identical modules.

2.2.2.3 Channel Attention Module

The preceding two modules increase the channel dimension of the original channel matrix from 4 to C and its multiples. While this facilitates the extraction of high-level channel features, it introduces feature ambiguity—making it difficult to distinguish which feature components contribute more significantly versus those with lesser contributions. Inspired by the channel attention mechanism, PCENet incorporates a channel attention module after the FFEM. The module computes importance weights for each channel in the high-dimensional feature matrix $\mathbf{H}_{FF}$ and adjusts feature channels proportionally based on the learned weight matrix.

The matrix $\mathbf{H}_{FF}$ is fed into the channel attention module, yielding a channel attention matrix $\mathbf{H}_{CA}$ with unchanged dimensions:

$$\mathbf{H}_{CA} = f_{CAM}(\mathbf{H}_{FF}), \tag{2.3}$$

where f_{CAM} denotes the channel attention module.

The network structure of the channel attention module is shown in Fig. 2.5. The average pooling layer performs channel-wise average pooling on the refined feature matrix $\mathbf{H}_{FF}$, generating pooled values for each channel. These values are then transformed through two Fully Connected (FC) layers to compute channel importance weights, where higher numerical values indicate greater importance. Finally, $\mathbf{H}_{FF}$

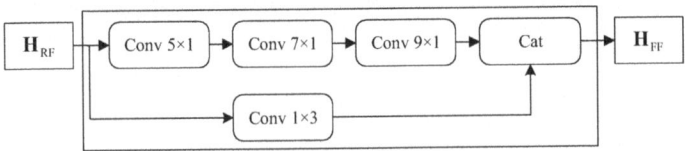

Fig. 2.4 FFEM structure

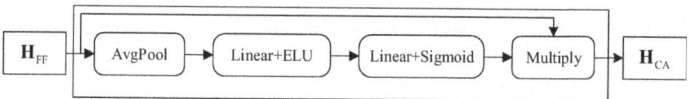

Fig. 2.5 CAM structure

is multiplied by the channel importance matrix to output the channel-attention-enhanced matrix $\mathbf{H}_{CA}$. Unlike the conventional channel attention mechanism, the module in PCENet replaces the activation function in the first FC layer with the Exponential Linear Unit (ELU) function, which better facilitates learning nonlinear relationships in channel data.

2.2.2.4 1 × 1 Convolutional Layer

Since the pilot-position CSI matrix $\mathbf{H}_{Pilot}$ has dimensions $M_P \times N_P \times 2$, while the channel attention matrix $\mathbf{H}_{CA}$ has dimensions $M_P \times N_P \times C$, a 1×1 convolutional layer is inserted before output $\mathbf{H}_{Pilot}$ to reduce the channel dimension of $\mathbf{H}_{CA}$ from C to 2. The 1×1 convolutional kernel is specifically chosen to minimize computational complexity.

The channel attention matrix $\mathbf{H}_{CA}$ is processed through this 1×1 convolutional layer to obtain the pilot-position CSI matrix:

$$\mathbf{H}_{Pilot} = \text{Conv}_{1\times 1}(\mathbf{H}_{CA}). \tag{2.4}$$

2.2.3 Loss Function for Channel Estimation

This chapter employs the Normalized Mean Square Error (NMSE) as both the model loss function and the evaluation metric for the discrepancy between estimated CSI and ground-truth CSI. For PCENet, the NMSE is defined as

$$L_{PCENet} = \frac{1}{N} \sum_{n=1}^{N} \frac{\sum_{i=1}^{M_P} \sum_{j=1}^{N_P} \left\| h_{n,ij} - \hat{h}_{n,ij} \right\|_2^2}{\sum_{i=1}^{M_P} \sum_{j=1}^{N_P} \left\| h_{n,ij} \right\|_2^2}, \tag{2.5}$$

where N represents the number of samples, M_P and N_P denote the quantities of subcarriers and OFDM symbols in the pilot pattern, $h_{n,i,j}$ corresponds to the actual CSI of the temporal-frequency grid corresponding to the ith subcarrier and jth OFDM symbol in the nth sample, $\hat{h}_{n,i,j}$ is the estimated $h_{n,i,j}$, and $\|\cdot\|_2$ indicates the L2-norm.

2.3 FCN-Based Channel Interpolation

2.3.1 Overall Network Structure Design

After obtaining the CSI at pilot positions, interpolation is required to estimate CSI at non-pilot positions, thereby acquiring full temporal-frequency-domain channel responses. Since the pilot estimation network in the previous section adopts a CNN-based structure, to avoid performance degradation caused by stacking identical network structures, this section employs FCN for deep interpolation. Inspired by the Multi-Layer Perceptron (MLP)-Mixer proposed in literature [23], we design a full temporal-frequency-domain deep channel INet based on FCN. This network learns the implicit mapping relationships between pilot and non-pilot positions while preserving the spatial characteristics of the channel matrix, ultimately yielding full temporal-frequency-domain CSI.

Figure 2.6 illustrates the overall structure of INet. It uses fully connected layers as the primary processing units and includes four sequentially connected modules tailored for channel interpolation: the Previous Up-sampling Module (PUM), Pre-patch Module (PM), Mixer Module (MM), and Channel Reconstruction Module (CRM).

2.3.2 Submodule Network Structure Design

2.3.2.1 Previous Upsampling Module

The input to INet is the CSI matrix $\mathbf{H}_{\text{Pilot}}$ at pilot positions, with dimensions $M_{\text{P}} \times N_{\text{P}} \times 2$. $\mathbf{H}_{\text{Pilot}}$ first undergoes the previous upsampling module, which applies bilinear interpolation sequentially along the subcarrier and OFDM symbol dimensions to upscale $\mathbf{H}_{\text{Pilot}}$ to the full temporal-frequency-domain size $M \times N \times 2$ (where M and N represent the number of subcarriers and OFDM symbols in the full temporal-frequency domain). The output is the previous upsampled matrix:

$$\mathbf{H}_{\text{PU}} = f_{\text{PUM}}(\mathbf{H}_{\text{Pilot}}), \tag{2.6}$$

where $f_{\text{PUM}}(\cdot)$ denotes the previous upsampling module.

Fig. 2.6 The architecture of INet

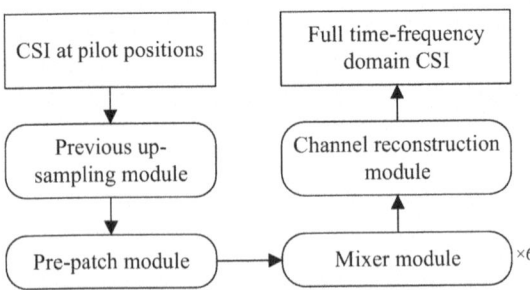

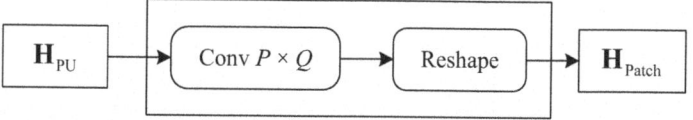

Fig. 2.7 The structure of the pre-patch module

2.3.2.2 Pre-patch Module

To extract spatial information of the channel using FCN, the pre-upsampled matrix $\mathbf{H}_{PU}$ is divided into multiple patches in the pre-patch module. Assuming each patch has a size of $P \times Q$ (where N/P and M/Q are integers), the output grouped matrix $\mathbf{H}_{Patch}$ has dimensions $(MN/PQ) \times C$. This process is expressed as

$$\mathbf{H}_{Patch} = f_{PM}(\mathbf{H}_{PU}), \qquad (2.7)$$

where $f_{PM}(\cdot)$ denotes the pre-patch module. C similarly performs upscaling on the channel dimension of the matrix.

The specific network structure of the pre-patch module is shown in Fig. 2.7. Unlike the pre-patch layer in MLP-Mixer, pre-patch module in INet utilizes convolutional layers for grouping, with both kernel size and stride set to $P \times Q$, and output channels set to C. This means that sub-matrices within each $P \times Q$ receptive field are grouped, resulting in an intermediate matrix of size $(M/P) \times (N/Q) \times C$. Subsequently, a dimensionality transformation is performed by merging the first two dimensions into a single dimension, ultimately outputting the grouped matrix $\mathbf{H}_{Patch}$ with dimensions $(MN/PQ) \times C$.

2.3.2.3 Mixer Module

The mixer module is the core of INet. It utilizes FCN to learn features along the horizontal and vertical dimensions of $\mathbf{H}_{Patch}$ and learn features along the subcarrier and OFDM symbol dimensions via matrix transposition, enabling full extraction of subcarrier and symbol correlations.

The mixer module processes $\mathbf{H}_{Patch}$ and outputs a dimension-preserved Mixer matrix:

$$\mathbf{H}_{Mixer} = f_{Mixer}(\mathbf{H}_{Patch}), \qquad (2.8)$$

where $f_{Mixer}(\cdot)$ denotes the Mixer module.

Figure 2.8 shows the structure of the Mixer module. $\mathbf{H}_{Patch}$ undergoes layer normalization and transposition, followed by a fully connected layer to extract temporal symbol correlations. After restoring the original shape via transposition

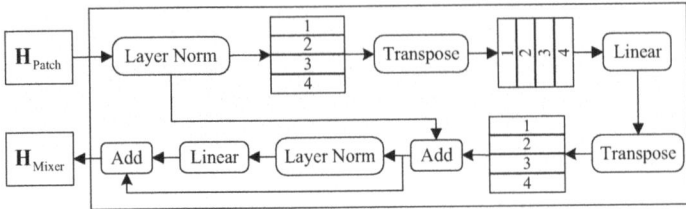

Fig. 2.8 The structure of the Mixer module

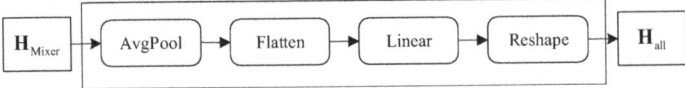

Fig. 2.9 The structure of the channel reconstruction module

and layer normalization, another fully connected layer extracts frequency subcarrier correlations. Two lateral connections enhance interpolation performance.

2.3.2.4 Channel Reconstruction Module

Since the dimensions of the Mixer matrix $\mathbf{H}_{\mathrm{mixer}}$ do not match those of the full temporal-frequency domain, a dimensionality transformation is required for $\mathbf{H}_{\mathrm{mixer}}$. Therefore, at the final stage of INet, the CRM is designed to transform the dimensions of $\mathbf{H}_{\mathrm{mixer}}$, yielding the full temporal-frequency-domain CSI matrix with dimensions $M \times N \times 2$:

$$\mathbf{H}_{\mathrm{all}} = f_{\mathrm{CRM}}(\mathbf{H}_{\mathrm{Mixer}}), \tag{2.9}$$

where $f_{\mathrm{CRM}}(\cdot)$ denotes the channel reconstruction module.

Figure 2.9 illustrates the detailed network structure of the channel reconstruction module. $\mathbf{H}_{\mathrm{Mixer}}$ first undergoes feature aggregation via an average pooling layer. The aggregated feature map is then flattened into a one-dimensional vector. Finally, the full temporal-frequency-domain CSI matrix $\mathbf{H}_{\mathrm{all}}$ is obtained through an FC layer and dimensionality transformation.

2.3.3 Loss Function for Channel Interpolation

INet uses NMSE as the loss function:

$$L_{\mathrm{INet}} = \frac{1}{N} \sum_{n=1}^{N} \frac{\sum_{i=1}^{M} \sum_{j=1}^{N} \|h_{n,ij} - \hat{h}_{n,ij}\|_2^2}{\sum_{i=1}^{M} \sum_{j=1}^{N} \|h_{n,ij}\|_2^2}, \tag{2.10}$$

where M and N are the subcarrier and OFDM symbol counts in the full temporal-frequency domain, and other variables align with (2.5).

2.4 Performance Evaluation

This section evaluates the performance of the pilot-position deep estimation network PCENet and the full temporal-frequency-domain deep interpolation network INet based on the channel dataset. First, the channel dataset used and the relevant model training parameters are introduced. Subsequently, the performances of PCENet and INet are tested and analyzed.

2.4.1 Dataset and Model Parameters

2.4.1.1 Dataset Generation

The channel dataset used in this chapter is generated by a standard link-level simulation platform employing Clustered Delay Line (CDL) channels. These channels include five types: CDL-A, CDL-B, CDL-C, CDL-D, and CDL-E. Among them, CDL-A, CDL-B, and CDL-C correspond to Non-Line of Sight (NLOS) transmission, while CDL-D and CDL-E are categorized as Line of Sight (LOS) transmission. The simulation parameters for generating the channel data are listed in Table 2.1.

The dataset contains a total of 200,000 channel samples, with 40,000 samples for each channel type. Additionally, each channel type includes data for five SNRs: 0 dB, 5 dB, 10 dB, 15 dB, and 20 dB, with 8000 samples per SNR. The 200,000 channel samples are divided into training, validation, and test sets in an 8:1:1 ratio,

Table 2.1 Simulation channel parameters

Parameter	Value
Channel name	Physical Downlink Shared Channel (PDSCH)
Transmit antennas	1
Receive antennas	1
Carrier frequency	0.92 GHz
Channel bandwidth	5 MHz
Duplex mode	TDD
Signal-to-Noise Ratio (SNR) (dB)	[0, 5, 10, 15, 20]
Subcarrier spacing	15 kHz
Channel coding	Low Density Parity Check (LDPC)
Cyclic Prefix (CP) Type	Normal CP
Modulation	Quadrature Phase Shift Keying (QPSK)

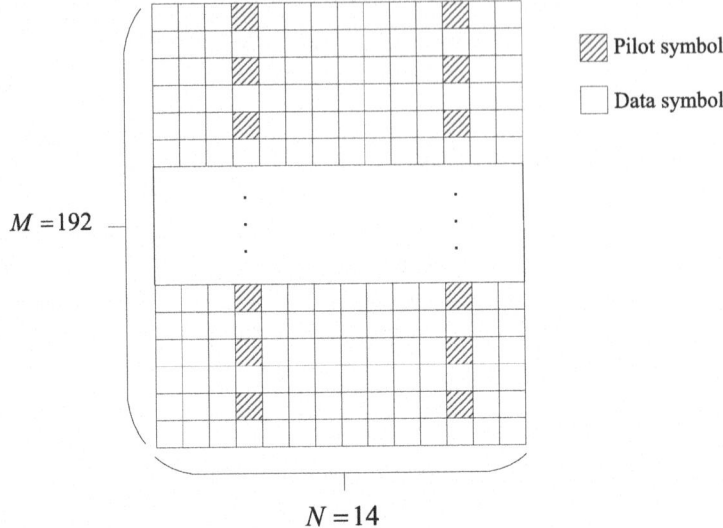

Fig. 2.10 Pilot pattern

resulting in 160,000 training samples, 20,000 validation samples, and 20,000 test samples. Furthermore, each time the network is trained, the training and validation sets are shuffled based on randomly generated index sequences to ensure different datasets for each training session.

In practical communication scenarios, channel SNR varies dynamically, and channel modeling cannot encompass all characteristics of real-world environments. To enhance the model's generalization capability and ensure applicability across diverse channel types and SNRs, the training and validation sets are uniformly mixed with the five channel types and five SNR levels. For comparison with traditional channel estimation algorithms, the test set separates data by different SNRs, with each SNR tested individually, while the channel types remain uniformly mixed.

The pilot pattern used in the simulation is shown in Fig. 2.10. It consists of 16 Resource Block (RB)s, including 192 subcarriers and 14 OFDM symbols, i.e., $M = 192$, $N = 14$. The pilot symbols occupy all odd-numbered subcarriers in the frequency domain and the 4th and 12th OFDM symbols in the time domain, i.e., $M_P = 96$, $N_P = 2$.

2.4.1.2 Model Training Parameters

In model training, the relevant parameters for training the PCENet and INet models are listed in Table 2.2. Notably, the optimizers used during training are not limited to a single type but adopt a combination of Adam and Stochastic Gradient Descent (SGD). Specifically, Adam is utilized for the first 300 epochs, followed by SGD for

Table 2.2 Parameters and
values of PCENet and INet

Parameter	Value
Pilot pattern size	96 × 2
Full temporal-frequency-domain size	192 × 14
Training samples	160,000
Validation samples	20,000
Test samples	20,000
Loss function	NMSE
Optimizer	Adam and SGD
Initial learning rate	0.001
Batch size	32
Epoch	500

the remaining 200 epochs. This hybrid optimization strategy helps further reduce the model's training error.

2.4.2 Performance of Channel Estimation

This subsection evaluates the channel estimation performance of PCENet at pilot positions, with comparative analysis against LS, practical LMMSE, and ideal LMMSE algorithms. PCENet takes the transmitted and received signals at pilot positions as the input and outputs CSI matrices at corresponding pilot positions; moreover, the ideal pilot-based channels are served as training labels.

Figure 2.11 illustrates the convergence performance of PCENet during training. The training process comprises two phases: The first 300 epochs use the Adam optimizer, both training and validation errors gradually decrease until reaching convergence; subsequently, the SGD optimizer is employed for the remaining 200 epochs, where the errors further decrease and eventually stabilize into a converged state.

Figure 2.12 compares the channel estimation performance of PCENet at pilot positions with LS, practical LMMSE, and ideal LMMSE algorithms. The estimation accuracy of all algorithms improves with increasing SNR from the figure, as higher SNR enhances the signal quality and consequently refines CSI estimation at pilot positions. Then, the performance of PCENet is compared with traditional algorithms. PCENet demonstrates significantly lower estimation errors than LS and practical LMMSE methods across all SNRs, confirming its substantial superiority over these methods; and when against with the ideal LMMSE, PCENet achieves lower estimation errors at 0 dB, 5 dB, and 10 dB, aligns with the ideal LMMSE at 15 dB, and exhibits marginally higher errors at 20 dB. This indicates PCENet's enhanced performance over ideal LMMSE at low SNRs and slightly inferior performance at high SNRs. However, the ideal LMMSE needs to obtain the delay power spectrum of the channel in advance, which is impossible to achieve in the actual communication systems. Therefore, considering both the estimation accuracy

Fig. 2.11 The convergence
of PCENet

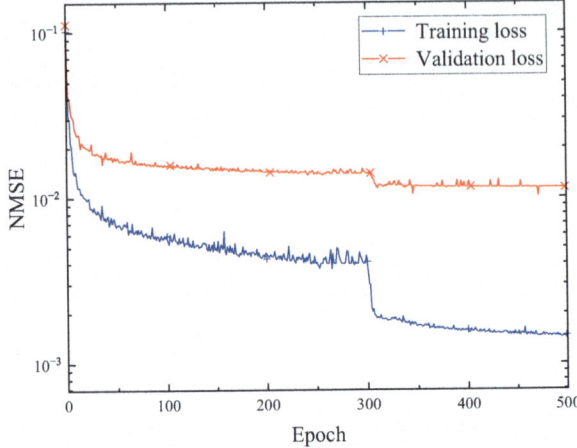

Fig. 2.12 Performance
comparison between PCENet
and traditional channel
estimation algorithms

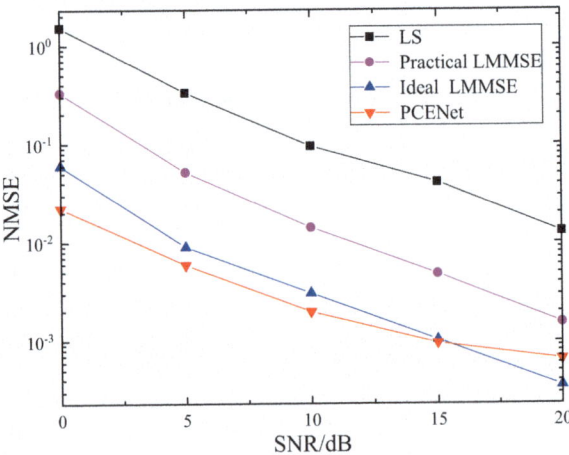

and implementation feasibility, PCENet can be considered as a more promising
and practical channel estimation algorithm at pilot positions compared to LS and
LMMSE.

To validate the generalization capability of PCENet across various channel types,
this subsection employs the NLOS CDL-A channel for the training set and the LOS
CDL-D channel for the test set to evaluate the model's estimation accuracy. The
results are compared with scenarios where both training and test sets utilize the
CDL-D channel, as illustrated in Fig. 2.13. It reveals that the model trained under the
CDL-D channel outperforms the model trained under the CDL-A channel across all
SNRs, because the NLOS environment (CDL-A channel) inherently exhibits poorer
channel conditions than LOS environment (CDL-D channel), yielding noisier and
lower quality training data. On the other hand, training under CDL-A channel fails

Fig. 2.13 Generalization of PCENet

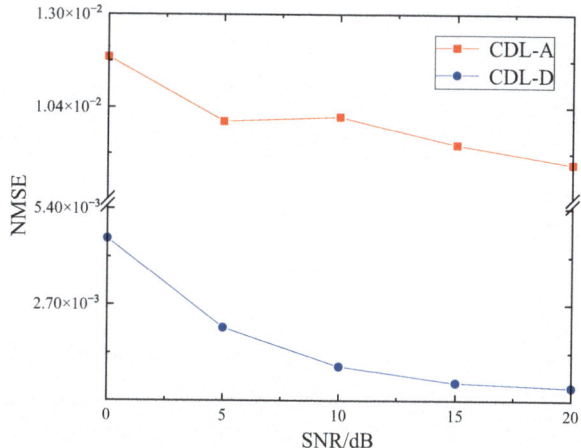

to adequately capture the statistical characteristics of CDL-D channel, highlighting PCENet's limited generalization capability across distinct channel types.

2.4.3 Performance of Channel Interpolation

This subsection evaluates the full temporal-frequency-domain interpolation performance of INet and compares it with the bilinear interpolation algorithm. Since the dataset in this study contains three traditional pilot-position channel estimation algorithms, and there exists three datasets of pilot-position CSI samples estimated by these conventional methods, during the training of INet, each dataset of pilot-position CSI data serves as input and outputs the corresponding full temporal-frequency-domain CSI, with the training labels of the ideal full temporal-frequency-domain channel responses. Subsequently, the performance of INet is systematically compared with that of the bilinear interpolation algorithm when both employ the three pilot-position CSI datasets as input.

Figure 2.14 presents the model convergence curves of INet trained using pilot-position CSI samples generated by practical LMMSE estimation. As shown in Fig. 2.14, INet exhibits rapid error reduction during initial training stages, subsequently enters a fluctuating phase, and ultimately achieves convergence with minor residual oscillations. The convergence pattern reveals that compared with CNN-based architecture, FCN-structured network demonstrates accelerated error descent during training while exhibiting greater error oscillations. Notably, during the training process of INet, the SGD optimizer cannot yield further error reduction, leading to the exclusive adoption of the Adam optimizer for training.

Figure 2.15 presents the performance comparison between INet and bilinear interpolation. From the figure, when pilot estimation employs LS and practical

Fig. 2.14 The convergence of INet

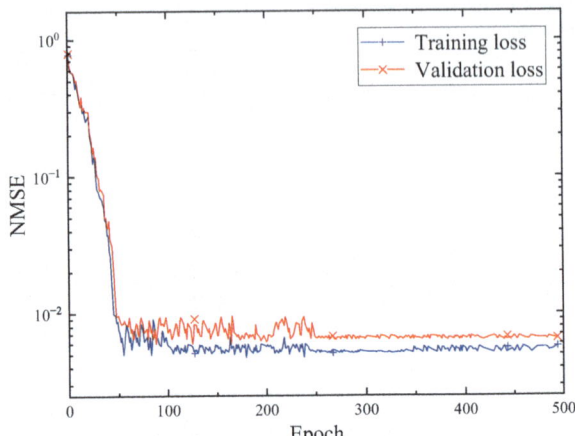

Fig. 2.15 Performance comparison between INet and bilinear interpolation

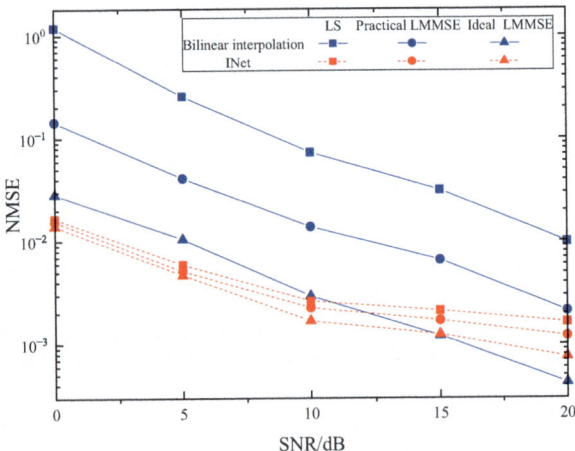

LMMSE algorithms, INet achieves lower NMSE for the interpolation compared to bilinear interpolation, demonstrating the superior performance of INet. However, with ideal LMMSE-based pilot estimation, INet exhibits lower interpolation errors than bilinear interpolation at 0 dB, 5 dB, and 10 dB SNRs, comparable performance at 15 dB, and marginally higher errors at 20 dB. This indicates that INet outperforms conventional bilinear interpolation at low SNRs but shows slightly diminished advantages at high SNRs. The subsequent analysis compares INet's performance across the three pilot estimation algorithms. Three red curves reveal a distinct performance hierarchy: Ideal LMMSE achieves the most optimal interpolation accuracy, followed by practical LMMSE, with LS yielding the poorest results. Notably, this performance disparity amplifies with increasing the SNR. This hierarchy aligns with theoretical expectations because that ideal LMMSE incorporates maximal channel prior knowledge (including second-order statistics, noise variance,

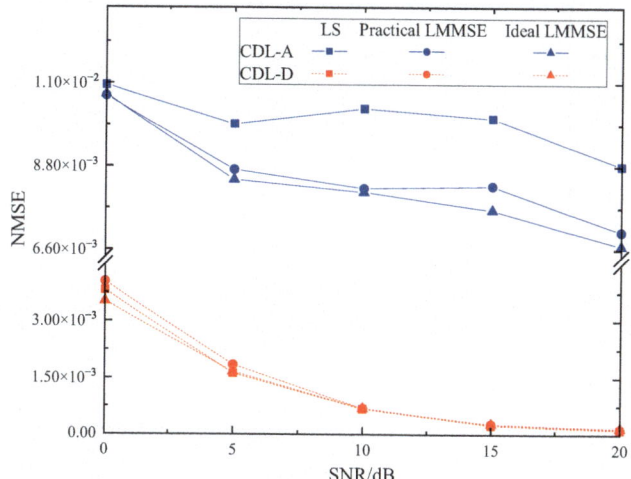

Fig. 2.16 Generalization of INet

and delay power profile), and generates pilot-position CSI closest to the ground truth, thereby enabling superior interpolation. Conversely, LS algorithm's simplistic estimation mechanism results in degraded pilot recovery and consequently inferior interpolation. Practical LMMSE balances algorithmic complexity and prior knowledge utilization and exhibits intermediate performance consistent with its theoretical foundations.

To validate the generalization capability of INet across various channel types similar to the PCENet, this subsection employs CDL-A channel for model training while evaluating interpolation performance on CDL-D channel. Comparative analysis is conducted against interpolation results obtained from models trained under CDL-D channel. Given the three conventional pilot-position estimation algorithms, the INet is trained separately using pilot-position CSI matrices generated by LS, practical LMMSE, and ideal LMMSE algorithms under both channel types, with ideal full temporal-frequency CSI matrices serving as training labels. Figure 2.16 demonstrates the channel generalization evaluation of INet. The results reveal that for all three pilot estimation algorithms, INet models trained under CDL-A channel exhibit degraded interpolation performance across all SNRs compared to the CDL-D trained counterparts, indicating limited channel generalization capacity. To alleviate the insufficient generalization ability of PCENet and INet for different channel types, it is necessary to make the training sample numbers of different channel types relatively average.

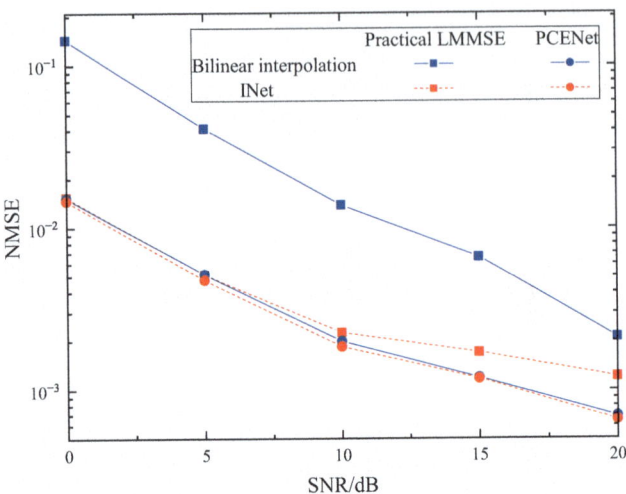

Fig. 2.17 Joint performance comparison of different channel estimation and interpolation methods

2.4.4 Performance of Two-Stage Channel Estimation

The channel estimation can be categorized into four distinct operational paradigms based on pilot-position estimation and full temporal-frequency interpolation methodologies: (i) conventional estimation + conventional interpolation, (ii) conventional estimation + DL-based interpolation, (iii) DL-based estimation + conventional interpolation, and (iv) DL-based estimation + DL-based interpolation. Given the suboptimal pilot estimation performance of LS algorithms and the impracticality of ideal LMMSE requiring prior knowledge of channel delay power profiles in the real world, this subsection adopts practical LMMSE as a balanced pilot estimation algorithm for the joint estimation capability of PCENet and INet.

Figure 2.17 illustrates the combined performance comparison between the practical LMMSE and PCENet, bilinear interpolation, and INet. It can be seen that the conventional estimation + conventional interpolation demonstrates the worst estimation accuracy with significant performance gaps compared to other combinations, substantiating the advantage of DL method in either estimation or interpolation stage. Then, the remaining three combinations exhibit comparable performance at 0–10 dB SNRs, and the conventional estimation + DL-based interpolation shows pronounced accuracy degradation at 15–20 dB SNRs, which indicates superior benefits from DL-based pilot estimation over DL-based interpolation. Finally, the DL-based estimation + DL-based interpolation method achieves marginally superior performance against DL-based estimation + conventional interpolation method, conclusively validating the outperformance of DL-based methodologies over conventional approaches throughout the channel estimation workflow.

2.5 Summary

This chapter proposes a two-stage deep channel estimation scheme based on traditional channel estimation algorithms, including the pilot-position channel estimation network PCENet and the full temporal-frequency-domain channel interpolation network INet. Firstly, considering the frequency-domain channel response as a two-dimensional image, the CNN-based PCENet is designed by leveraging the powerful feature extraction capability of CNN structure. The PCENet takes the transceiver signals at pilot positions as input and outputs the CSI matrix at these pilot locations. Subsequently, treating the pilot-position CSI matrix as a noisy low-resolution image, the FCN-based INet is designed to utilize the dimensional transformation capability of FCN structure. The INet performs interpolation on the pilot-position CSI matrix by exploiting the correlation characteristics between subcarriers and OFDM symbols and then ultimately obtains the full temporal-frequency-domain CSI. Finally, comprehensive simulations and analysis are conducted by using the CDL channel dataset to validate the proposed models. Initially, a comparative analysis is performed between PCENet and conventional channel estimation algorithms to validate PCENet's superior performance for the pilot-based channel estimation. Then, the interpolation error is compared between the INet and the traditional bilinear interpolation algorithm across three classical pilot estimation schemes, concluding that the INet demonstrates the advantage over conventional interpolation methods. Finally, a joint performance analysis integrating PCENet and INet is implemented, with comparative assessments against practical LMMSE and bilinear interpolation. The experimental results reveal that DL-based approaches can indeed achieve measurable performance enhancements for the channel estimation.

References

1. He, H., Jin, S., Wen, C. et al.: Model-driven deep learning for physical layer communications. IEEE Wirel. Commun. **26**(5), 77–83 (2019)
2. Yan, M., Feng, G., Zhou, J., et al.: Intelligent resource scheduling for 5G radio access network slicing. IEEE Trans. Veh. Technol. **68**(8), 7691–7703 (2019)
3. Xie, R., Jia, X., Wu, K.: Adaptive online decision method for initial congestion window in 5G mobile edge computing using deep reinforcement learning. IEEE J. Sel. Areas Commun. **38**(2), 389–403 (2020)
4. Xie, H., Qin, Z., Li, G., et al.: Deep learning enabled semantic communication systems. IEEE Trans. Signal Process. **69**(1), 2663–2675 (2021)
5. Beek, J., Edfors, O., Sandell, M., et al.: On channel estimation in OFDM system. In: IEEE 45th Vehicular Technology Conference, pp. 815–819 (1995)
6. Jones, V., Raleigh, G.: Channel estimation for wireless OFDM systems. In: IEEE GLOBECOM, pp. 980–985 (1998)
7. Dang, J., Cheng, L., Yuan, M.: Research on least square channel estimation algorithm based on phase compensation. In: 2nd IEEE International Conference on Computer and Communications (ICCC), pp. 1647–1650 (2016)

8. Wang, H., Zhou, F., Yu, P., et al.: An improved threshold wavelet denoising LS channel estimation algorithm based on IoT. In: International Wireless Communications and Mobile Computing (IWCMC), pp. 326–330 (2020)
9. Hasan, A., Motakabber, S., Anwar, F., et al.: A computationally efficient least squares channel estimation method for MIMO-OFDM systems. In: 8th International Conference on Computer and Communication Engineering (ICCCE), pp. 331–334 (2021)
10. Suyan, N., Saini, G.: Low complexity MMSE channel estimator for downlink MC-CDMA system. In: 2nd International Conference on Inventive Systems and Control (ICISC), pp. 706–709 (2018)
11. Almamori, A., Mohan, S.: Improved MMSE channel estimation in massive MIMO system with a method for the prediction of channel correlation matrix. In: IEEE 8th Annual Computing and Communication Workshop and Conference (CCWC), pp. 670–672 (2018)
12. Deng, Y., Ohtsuki, T.: Low-complexity subspace MMSE channel estimation in massive MU-MIMO system. IEEE Access 8, 124371–124381 (2020)
13. Ye, H., Li, G., Juang, B.: Power of deep learning for channel estimation and signal detection in OFDM systems. IEEE Wirel. Commun. Lett. 7(1), 114–117 (2018)
14. Gao, X., Jin, S., Wen, C., et al.: ComNet: Combination of deep learning and expert knowledge in OFDM receivers. IEEE Commun. Lett. 22(12), 2627–2630 (2018)
15. He, H., Wen, C., Jin, S., et al.: Deep learning-based channel estimation for beamspace mmWave massive MIMO systems. IEEE Wirel. Commun. Lett. 7(5), 852–855 (2018)
16. Zhang, K., Zuo, W., Chen, Y., et al.: Beyond a Gaussian denoiser: residual learning of deep CNN for image denoising. IEEE Trans. Image Process. 26(7), 3142–3155 (2017)
17. Sun, H., Zhao, Z., Fu, X., et al.: Limited feedback double directional massive MIMO channel estimation: From low-rank modeling to deep learning. In: IEEE 19th International Workshop on Signal Processing Advances in Wireless Communications (SPAWC), pp. 1–5 (2018)
18. Soltani, M., Pourahmadi, V., Mirzaei, A., et al.: Deep learning-based channel estimation. IEEE Commun. Lett. 23(4), 652–655 (2019)
19. Dong, C., Loy, C., He, K., et al.: Image super-resolution using deep convolutional networks. IEEE Trans. Pattern Anal. Mach. Intell. 38(2), 295–307 (2016)
20. Dong, P., Zhang, H., Li, G., et al.: Deep CNN for wideband mmWave massive MIMO channel estimation using frequency correlation. In: ICASSP 2019—2019 IEEE International Conference on Acoustics, Speech and Signal Processing (ICASSP), pp. 4529–4533 (2019)
21. Moon, S., Kim, H., Hwang, I.: Deep learning-based channel estimation and tracking for millimeter-wave vehicular communications. J. Commun. Netw. 22(3), 177–184 (2020)
22. Pan, J., Shan, H., Li, R., et al.: Channel estimation based on deep learning in vehicle-to-everything environments. IEEE Commun. Lett. 25(6), 1891–1895 (2021)
23. Tolstikhin, I., Houlsby, N., Kolesnikov, A., et al.: MLP-mixer: an all-MLP architecture for vision (2021). https://arxiv.org/abs/2105.01601

Chapter 3
Intelligent CSI Feedback Technology for FDD Systems

Abstract DL-based CSI feedback has the potential to improve the recovery accuracy and reduce the feedback overhead in massive MIMO-OFDM systems. This chapter first focuses on the eigenvector-based CSI feedback and designs a DL-based lightweight network, referred to as MixerNet, where the joint eigenvector composed of multiple subbands is compressed at the UE and then recovered at the gNB. On the other hand, an adaptive bidirectional long short-term memory network (ABLNet) for CSI feedback is designed to process various input CSI lengths. Then, to realize a more flexible feedback bit number, a feedback bit control unit (FBCU) module is proposed to control the output length of feedback bits. Based on which, a target feedback performance can be adaptively achieved by a designed bit number adjusting (BNA) algorithm.

Keywords AI · MIMO · OFDM · FDD · Eigenvector · CSI feedback

3.1 Background

MIMO technology has become the key technology of 5G mobile communication due to its high spectrum efficiency [1, 2]. To generate beamforming for downlink signal transmission, it is necessary to obtain timely and accurate downlink CSI at the gNB. In TDD mode, the downlink CSI can be directly estimated from uplink pilot by using channel reciprocity [3], while in FDD mode, different frequency bands are employed for uplink and downlink, making it challenging to estimate the downlink CSI from uplink pilot. Consequently, the CSI feedback scheme for massive MIMO in FDD systems has become one of the hot topics in recent years.

The traditional CSI feedback methods, such as Compressive Sensing (CS) algorithms [4, 5], utilize the channel sparsity and orthogonal basis matrix to compress CSI and then feedback the low-dimensional CSI to the gNB for reconstruction. However, the actual channel is not completely sparse on any orthogonal basis. Moreover, both the number of algorithm iterations and the feedback overhead increase dramatically with the increasing number of antennas in massive MIMO

L. Zhao et al., *AI for Wireless Physical Layer*, SpringerBriefs in Computer Science, https://doi.org/10.1007/978-3-032-01367-5_3

systems. Therefore, feasible CSI feedback methods for massive MIMO systems need to be further studied.

Recently, AI has been developed rapidly, including DL [6–8] and Reinforcement Learning (RL) technology [9–11]. In particular, wireless communication systems based on DL frameworks have attracted a lot of attention, which can efficiently achieve channel estimation [12, 13], feedback [14–21], channel encoding[22], and signal detection [23]. Specifically, the DL-based CSI feedback, including full channel CSI feedback and eigenvector-based CSI feedback, has been widely studied since it can provide higher recovery accuracy and lower feedback overhead simultaneously.

For the full channel CSI feedback, a kind of CSI feedback architecture based on the autoencoder structure [24], termed as CsiNet, was first introduced in [14]. Based on which, a series of feedback models were proposed to improve the feedback accuracy by optimizing the model structure or quantizing method of the compressed vector with continuous values [15–19]. On the other hand, for the eigenvector-based CSI feedback discussed by 3rd Generation Partnership Project (3GPP) [25], fully connected layers and CNN are adopted to design EVCsiNet [20]. Moreover, the encoder of Transformer was used for both the encoder and decoder of CSI feedback model to compress and recover the eigenvectors [21]. However, both these eigenvector-based CSI feedback models are more complex.

Additionally, the studies in [14–21] primarily focus on the design of DL-based models or quantization methods to improve the CSI feedback performance, and they need to train and store a large number of feedback models to handle different input CSI lengths and feedback bit numbers. Therefore, for the full channel CSI feedback, the models with adaptive inputs or Correlation Rotation (CR)s are studied. On the one hand, DL-based feedback schemes only relying on fully convolutional are proposed for CSI with different numbers of subcarriers and antennas; however, realizing multiple CRs needs a list of model parameters [26–28]. On the other hand, a framework consisting of an encoder with multiple fully connected layers and multiple decoders is proposed to realize various CRs by leveraging optional output layer and corresponding decoder in [29]; furthermore, its variant simplifies multiple decoders into a multibranch decoder [30]. However, both of them need more fully connected layers of encoder and decoders (or branches) to accommodate various CRs; meanwhile, the adaptive input model in [26–28] could not be compatible with the multi-CRs model in [29, 30]. Additionally, the padding operation is taken to change the lengths of compressed vectors [31]; a classification model is designed to select the suitable CR for the CSI data with same size in the training stage but cannot be flexibly adjusted in the inference stage [32]; a quantization method with adjusted quantization bit number of each neuron output value is proposed for variable CRs without considering the quantization loss in the test stage [33]. In conclusion, the proposed adaptive schemes may not be sufficient for alterable input CSI lengths and feedback bit numbers simultaneously with one set of model parameters.

Therefore, to address the overchallenges, this chapter first proposes a lightweight DL-based feedback model; then, an adaptive eigenvector-based CSI feedback model and a plug-in control unit are designed for adapting to various input CSI lengths and feedback bit numbers simultaneously; moreover, a Bit Number Adjusting (BNA) algorithm is developed for satisfying the target feedback performance. In summary, the main contributions of this chapter are given as follows:

- In order to reduce the complexity of the eigenvector-based CSI feedback model, a DL-based lightweight feedback model is first designed, named as MixerNet, which mainly consists of Mixer layers. Similarly to both EVCsiNet and Transformer, the joint eigenvector is first compressed at the encoder and then recovered at the decoder. In addition, two kinds of quantization method Uniform Quantization (UQ) and Vector Quantization (VQ) are studied after the output of compressed data at the encoder.

- To deal with different lengths of input CSI, an input-adaptive CSI feedback model based on autoencoder structure is designed by leveraging BiLSTM, referred to as ABLNet. By fully exploiting the property of LSTM for processing different lengths of input sequences, the proposed ABLNet is capable of compressing and recovering input CSI with different lengths, and the number of feedback bits is in proportion to the corresponding length of input CSI.

- Then, an Feedback Bit Control Unit (FBCU) is proposed to further realize the adjustable number of feedback bits through discarding the ending part of codeword output by the encoder. Based on the adjustable characteristic of the proposed ABLNet with FBCU, a BNA algorithm is designed to achieve a unified, instead of an average, target feedback performance for every input CSI with lower feedback bit number on average than before the adjustment.

3.2 System Model for CSI Feedback

This section firstly introduces massive MIMO-OFDM system model and then discusses the DL-based CSI feedback model in detail.

3.2.1 Massive MIMO-OFDM System Model

In a massive MIMO-OFDM system, N_T transmit antennas are deployed at the gNB, and N_R receive antennas are deployed at each UE. As illustrated in Fig. 3.1, by leveraging channel estimation at the UE [34], the downlink channels in frequency domain can be obtained based on the pilot sequences transmitted by the gNB. Then, the downlink channels are divided into K subbands, where each subband consists of N_{SC} subcarriers and the subcarriers in the same subband employ the same beamformer at the gNB in order to reduce the system complexity. The common beamformer in each subband can be obtained by the following method in order to maintain the performance of each subcarrier.

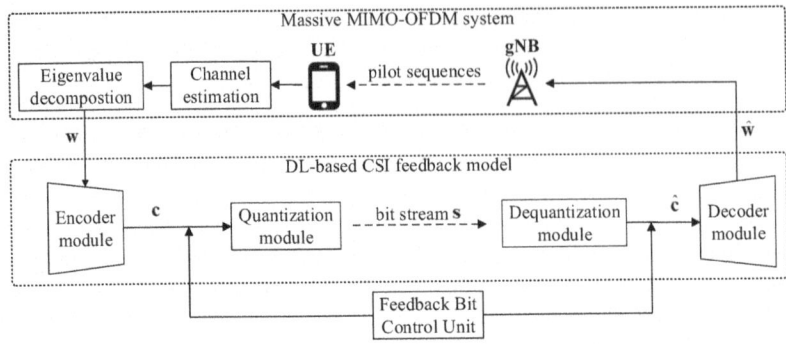

Fig. 3.1 The massive MIMO–OFDM system model and adaptive CSI feedback framework

Assume that the downlink channel of the nth subcarrier in the kth subband is denoted as $\mathbf{H}_{kn} \in \mathbb{C}^{N_R \times N_T}$ ($1 \leq k \leq K, 1 \leq n \leq N_{SC}$). Then, the correlation matrix of the channel of each subcarrier in the kth subband can be calculated as $\mathbf{H}_{kn}^H \mathbf{H}_{kn}$, and the average correlation matrix of the kth subband can be written as

$$\mathbf{R}_k = \frac{1}{N_{sc}} \sum_{n=1}^{N_{sc}} \mathbf{H}_{kn}^H \mathbf{H}_{kn}. \tag{3.1}$$

Based on Eigenvalue Decomposition (EVD), the feedback eigenvector of the kth subband can be calculated by

$$\mathbf{R}_k \mathbf{w}_k = \lambda_k \mathbf{w}_k, \tag{3.2}$$

where λ_k and $\mathbf{w}_k = \left[w_{k1}, w_{k2}, \cdots, w_{kN_T} \right]^T \in \mathbb{C}^{N_T \times 1}$ represent the maximum eigenvalue and corresponding eigenvector of matrix $\mathbf{R}_k$ in the kth subband. From (3.2), $\mathbf{w}_k$ is a complex vector; to fit for the processing of universal neural network, the corresponding real-valued eigenvector is given by

$$\tilde{\mathbf{w}}_k = \left[\mathrm{Re}\left\{ w_{k1} \right\}, \mathrm{Im}\left\{ w_{k1} \right\}, \cdots, \mathrm{Re}\left\{ w_{kN_T} \right\}, \mathrm{Im}\left\{ w_{kN_T} \right\} \right]^T, \tag{3.3}$$

where $\tilde{\mathbf{w}}_k \in \mathbb{R}^{2N_T \times 1}$ and $\mathrm{Re}\left\{ \cdot \right\}$ and $\mathrm{Im}\left\{ \cdot \right\}$ represent the real value and the imaginary value of w_{kn} ($1 \leq n \leq N_T$), respectively. Therefore, the joint real eigenvector $\mathbf{w}$ of the K subbands can be written as

$$\mathbf{w} = \left[\tilde{\mathbf{w}}_1, \tilde{\mathbf{w}}_2, \cdots, \tilde{\mathbf{w}}_K \right]^T \in \mathbb{R}^{K \times 2N_T}. \tag{3.4}$$

Moreover, CSI eigenvector with different lengths means that different CSI joint eigenvectors have a different number of subbands, i.e., K.

3.2.2 DL-Based CSI Feedback Model

Since the CSI feedback scheme is comparable to the overall workflow of the autoencoder, the autoencoder architecture can be employed to design the DL-based CSI feedback framework. As shown in Fig. 3.1, an adaptive CSI feedback model is proposed to adapt to the input CSI eigenvector with different lengths and various numbers of feedback bits.

Among the feedback model, the UE first obtains the CSI eigenvector $\mathbf{w}$ and compresses it to codeword vector $\mathbf{c}$ by the encoder, which is denoted as

$$\mathbf{c} = f_{\theta_\mathrm{E}}\left(\mathbf{w}, K\right), \tag{3.5}$$

where $f_{\theta_\mathrm{E}}\left(\cdot\right)$ is the encoder function with parameter set θ_E. And the length m of codeword vector $\mathbf{c}$ changes proportionally with K.

Then, an FBCU module could be applied to make the codeword vector $\mathbf{c}$ changeable based on the capacity requirements of feedback channel. If the number q of quantization bit is fixed, the length Q of feedback bits obtained after quantization is also changeable according to the length n ($\leq m$) of $\mathbf{c}$. Therefore, if the FBCU module is applied, (3.5) becomes

$$\mathbf{c} = f_{\theta_\mathrm{E}}\left(\mathbf{w}, K, n\right). \tag{3.6}$$

After passing through the quantization and dequantization modules, the obtained feedback bitstream $\mathbf{s}$ and recovered codeword vector $\hat{\mathbf{c}}$ can be respectively expressed as

$$\mathbf{s} = f_{\mathrm{quan}}\left(\mathbf{c}\right), \hat{\mathbf{c}} = f_{\mathrm{dequan}}\left(\mathbf{s}\right), \tag{3.7}$$

where $f_{\mathrm{quan}}\left(\cdot\right)$ and $f_{\mathrm{dequan}}\left(\cdot\right)$ are the quantization and dequantization functions, respectively.

Finally, the gNB reconstructs the CSI eigenvector $\hat{\mathbf{w}}$ with the decoder module, which is denoted as

$$\hat{\mathbf{w}} = f_{\theta_\mathrm{D}}\left(\hat{\mathbf{c}}, K\right), \tag{3.8}$$

where $f_{\theta_\mathrm{D}}\left(\cdot\right)$ is the decoder function with parameter set θ_D.

Usually, the Square of Generalized Cosine Similarity (SGCS) is taken to evaluate the recovery performance of eigenvector-based CSI feedback. The average SGCS ρ between the original joint eigenvector $\mathbf{w}$ and the recovered joint eigenvector $\hat{\mathbf{w}}$ with K subbands can be written as

$$\rho\left(\theta_\mathrm{E}, \theta_\mathrm{D}\right) = \frac{1}{K} \sum_{k=1}^{K} \left(\frac{\left\|\mathbf{w}_k^\mathrm{H} \hat{\mathbf{w}}_k\right\|}{\left\|\mathbf{w}_k\right\| \left\|\hat{\mathbf{w}}_k\right\|} \right)^2. \tag{3.9}$$

Then, the main objective of this chapter is to optimize the model weight parameter set $\boldsymbol{\Theta}_1 = \{\boldsymbol{\theta}_{\mathrm{E}}, \boldsymbol{\theta}_{\mathrm{D}}\}$ to make the feedback performance ρ as high as possible, which can be written as

$$\boldsymbol{\Theta}_1 = \arg\max_{\boldsymbol{\Theta}_1} \rho\left(\boldsymbol{\theta}_{\mathrm{E}}, \boldsymbol{\theta}_{\mathrm{D}}\right). \tag{3.10}$$

3.3 Lightweight MixerNet for CSI Feedback

3.3.1 Structure Design of MixerNet

The architecture of proposed MixerNet is shown in Fig. 3.2. When designing and building the model, the main consideration is to propose a high accuracy model with low computation complexity, so the encoder and decoder structures of the model are roughly the same, which also refers to the schemes of Transformer applied to CSI feedback [21].

Encoder As shown in the left column of Fig. 3.2, the encoder takes the joint eigenvector $\mathbf{w} \in \mathbb{R}^{1 \times (K \times 2N_T)}$ as model input. The input is first reshaped to a $K \times 2N_T$ two-dimensional matrix. Then, 36 Mixer layers are employed for data

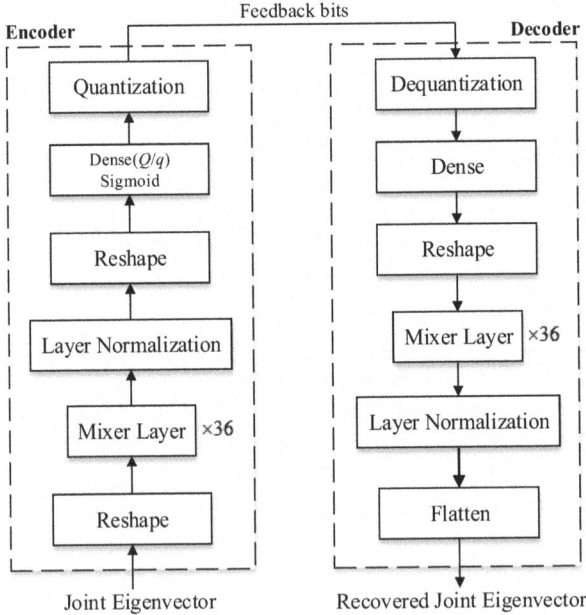

Fig. 3.2 MixerNet structure

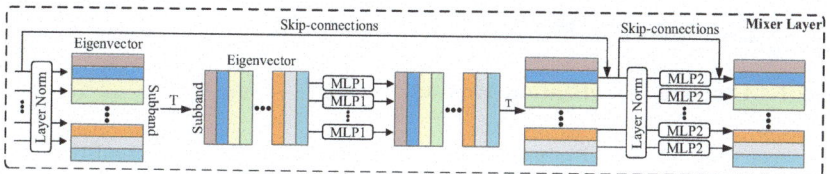

Fig. 3.3 Mixer layer structure

feature extraction, which is the core module in our proposed model and will be discussed below in detail. After 36 Mixer layers, the output is subjected to the layer normalization operation and is reshaped to $1 \times (K \times 2N_T)$. Next, a full connection layer with Q/q units is adopted, where Q is the length of feedback bitstream and q is the quantization bits for each float number. Meanwhile, the sigmoid activation function is used after the full connection layer to introduce nonlinearity. Moreover, the quantization layer is employed to transform the float vector with length Q/q into the bitstream output of the encoder with length Q. Different quantization methods have significant effect on the model performance and therefore will be described in the next subsection in detail.

Specially, the structure of a Mixer layer is illustrated in Fig. 3.3, which is similar to the MLP Mixer [35]. In a Mixer layer, we can notice that token mixing MLPs (MLP1) and channel mixing MLPs (MLP2) are used to map the columns and rows of the two-dimensional matrix, respectively. MLP1 allows the communication between different spatial locations, acting on each subband column and enabling spatial information mixing across subbands, while MLP2 allows communication between different channels, acting on different lines of the two-dimensional matrix and realizing the mixing of different subbands. MLP1 and MLP2 share weights in the mapping process in different columns and rows and are consistent with the fully connection layer structure. In addition, the skip connection and layer normalization are added to the Mixer layer to improve the model performance.

Decoder The structure of decoder is shown in the right column of Fig. 3.2. The dequantization layer is first used to transform the received bitstream into a float number vector, which is the inverse operation of the quantization layer at the encoder. Then, a full connection layer with $K \times 2N_T$ units is used and reshapes its output from $1 \times (K \times 2N_T)$ to $K \times 2N_T$, which is a two-dimensional matrix regarded as the input of Mixer layer. Next, 36 Mixer layers are employed to reconstruct the data feature with the layer normalization operation, and the structure of each Mixer layer is the same to the encoder. Finally, the data is flattened to recover the dimension of original joint eigenvector, which is regarded as the recovered joint eigenvector $\hat{\mathbf{w}}$ of the decoder.

3.3.2 Quantization Method

3.3.2.1 Uniform Quantization Method

The basic method of UQ is to quantize the input values by equidistant divisions. Based on this idea, the last quantization layer in encoder can adopt the UQ method. The quantization interval is determined by the quantization bit number q, and the quantization interval can be obtained as $[0, 2^q - 1]$. The schematic diagram of UQ process is shown in Fig. 3.4. The unquantized data is input into the quantization layer and quantized into $m_a, m_b, \cdots, m_{(Q/q)-1}, m_{Q/q}$ according to the quantization bit number q, where $m_a, m_b, \cdots, m_{(Q/q)-1}, m_{Q/q} \in [0, 2^q - 1]$ and are all positive integers. Then, the quantized results are concatenated and passed into the "Num2Bit" module to be converted into bitstream form, which will be dequantized and recovered at the decoder.

3.3.2.2 Vector Quantization Method

The basic method of VQ is to generate several scalar data in a vector and then quantize it in a given codebook space. The quantization method is to calculate the Euclidean distance between the vector to be quantized and the vector in the codebook and select the vector with the nearest distance as the output of quantization, so as to compress the data. Therefore, we adopt VQ method in the quantization layer referred to the VQ-VAE model [36].

Considering that float numbers with length Q/q are regarded as the whole vector for VQ, the codebook may be too large and will occupy a large amount of storage space and even make memory overflow. Therefore, we consider the form of grouping VQ, i.e., the float vector is divided into multiple sub-vectors and each sub-vector is denoted as $\mathbf{v}_i$ ($i \in \{1, 2, \cdots, n\}$). The group number n can be determined by the feedback bits Q and the number of vectors M in the codebook, which can be written as

$$n = \frac{Q}{\log_2 M}. \tag{3.11}$$

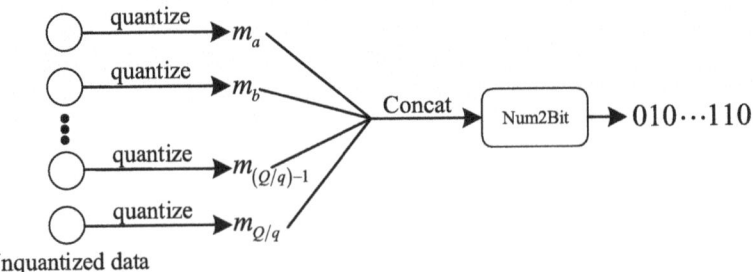

Fig. 3.4 UQ process

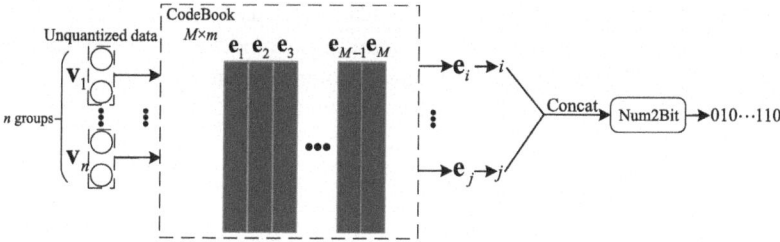

Fig. 3.5 VQ process

Meanwhile, the length m of each sub-vector $\mathbf{v}_i$ can be calculated as

$$m = \frac{Q}{qn} = \frac{\log_2 M}{q}. \tag{3.12}$$

Therefore, the size of codebook is M. In particular, the codebook $M \times m$ can be regarded as a two-dimensional weight matrix, which will also participate in training and updating.

Moreover, the schematic diagram of VQ process is shown in Fig. 3.5. Q/q unquantized float numbers are divided into n groups with m numbers in each group. Then, the Euclidean distance between the float sub-vector for each group and the vectors in codebook is calculated, and the vector with the nearest distance d_i ($i \in \{1, 2, \cdots, n\}$) is selected as the quantization result for each group of float sub-vector. Specifically, the nearest distance d_i can be written as

$$d_i = \|\mathbf{v}_i - \mathbf{e}_k\|^2, \tag{3.13}$$

where $\mathbf{e}_k$ is the nearest vector of the unquantized sub-vector v_i. Then, the indexes corresponding to the selected vectors are concatenated together and transmitted to the "Num2Bit" module to be converted into bitstream form. The bitstream is then passed to the decoder for dequantization and recovery.

3.3.2.3 Loss Function Design

Under the UQ method, the cosine similarity loss function is adopted as the loss function, which can be written as

$$L_C = 1 - \rho^2, \tag{3.14}$$

while under the VQ method, except the cosine similarity loss, the quantization loss is added to the loss function in order to improve the model performance and speed up the convergence of the model. The quantization loss can be given by

$$L_Q = \sum_{i=1}^{n} d_i^2 = \sum_{i=1}^{n} \|\mathbf{v}_i - \mathbf{e}_k\|^2. \tag{3.15}$$

Therefore, the total loss function could be described as

$$L = L_C + \beta \times L_Q, \tag{3.16}$$

where $\beta \in (0, 1]$ is a regularization coefficient to balance the proportion of the quantization loss in the total loss.

3.3.3 Performance Evaluation

In this section, the simulation results are provided to verify the advantage of the proposed MixerNet model compared with EVCsiNet model [20] and Transformer model [37].

3.3.3.1 Simulation Setup

In this chapter, we adopt the CDL channel model defined in 3GPP to generate the data samples, specially the CDL-C30 channel model with $K = 6$ subbands, and 30ns delay spread is employed in our simulation. Moreover, a typical setup that $N_T = 32$ transmit antennas at the gNB and $N_R = 4$ receive antennas at UE is adopted in the MIMO-OFDM system. Therefore, the size of CSI eigenvector is $K \times 2N_T = K \times 64$ ($K = 3, 6, 12$). Based on the CDL-C channel, 100,000 data samples are considered, as well as the training set, validation set, and test set have 98,010, 990, and 1000 samples, respectively. For the training phase, the default Adaptive Momentum (Adam) optimizer is employed with learning rate $l_r = 0.001$ and trains 500 epochs. Additionally, for the VQ loss function (3.16), we set $\beta = 1$.

3.3.3.2 Convergence Analysis

To verify the convergence of our proposed MixerNet model, the loss curves for UQ method and VQ method versus the number of training epochs are illustrated in Fig. 3.6a, b, respectively. In this chapter, we consider three cases that feedback bits Q are 32 bits, 48 bits, and 120 bits. From Fig. 3.6a, b, we can clearly notice that the loss values of both the training set and validation set drop rapidly in the early stage

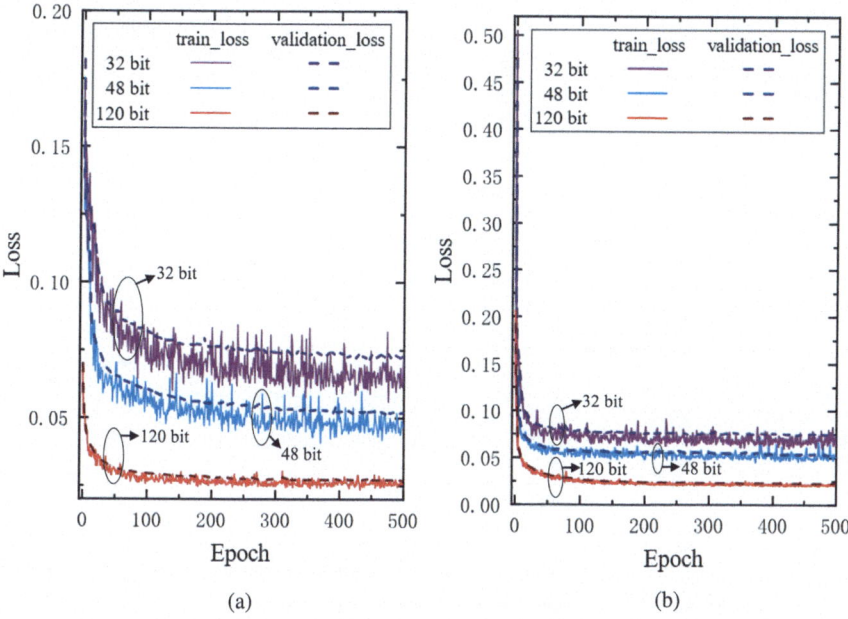

Fig. 3.6 Loss curves versus the number of training epochs. (**a**) UQ method. (**b**) VQ method

of training, and then downward trend becomes flat. In particular for the training set in Fig. 3.6a, the loss curve fluctuates with the increasing number of epochs when $Q = 32$ and $Q = 48$; however, the overall tendency is still stationary. Therefore, we can judge that the models under UQ method and VQ method converge. Moreover, no matter which quantization method is adopted, we can see that the larger feedback bits Q are, the smaller the loss values of training set and validation set are, which will result in better performance of the model.

3.3.3.3   Performance Comparison

The performance comparison between the proposed MixerNet and EVCsiNet and Transformer under both UQ and VQ is listed in Table 3.1.

For each model, whether the UQ or VQ for quantization method is adopted, the performance is gradually improved with the increase of the feedback bits Q, which is corresponded to the communication theory. Then, for both the EVCsiNet and MixerNet, the performance under VQ is better than that under UQ for the same feedback bits Q. This is consistent with our conjecture before the experiment, because the quantization result obtained by VQ is still a float number, not positive integer by UQ. Therefore, the quantization results under VQ as the input of the decoder can provide more data features and achieve the better recover of eigenvector. However, for the Transformer model, only when $Q = 120$, VQ

Table 3.1 Comparison of different models with different quantization methods

Model+ Quantization method	ρ^2 Performance		
	32 bits	48 bits	120 bits
EVCsiNet+UQ	0.913	0.934	0.968
EVCsiNet+VQ	0.926	0.945	0.973
Transformer+UQ	0.919	0.948	0.977
Transformer+VQ	0.850	0.924	0.978
MixerNet+UQ	0.929	0.949	0.973
MixerNet+VQ	0.935	0.954	0.978

Table 3.2 Trainable parameters and FLOPs of different models

Models	Trainable parameters ($\times 10^6$)			FLOPs ($\times 10^6$)		
	32 bits	48 bits	120 bits	32 bits	48 bits	120 bits
EVCsiNet	4.67	4.68	4.75	9.35	9.38	9.51
Transformer	21.42	21.43	21.50	42.85	42.87	43.00
MixerNet	1.20	1.22	1.27	2.41	2.46	2.55

method outperforms UQ method, while UQ method has better performance when $Q = 32$ and $Q = 48$. The reason may be that more information is lost by compression when the feedback bits Q are low, and the model with VQ method is more complicated, resulting in poor final performance. Moreover, under both UQ and VQ, the performance of proposed MixerNet is significantly improved compared with EVCsiNet and Transformer, except that when $Q = 120$ and UQ is adopted, Transformer has the same performance as MixerNet. Specifically, the proposed MixerNet can achieve 0.016, 0.015, and 0.005 performance gain on CSI recovery accuracy compared with EVCsiNet under UQ method. Therefore, the simulation results have demonstrated the superiority of our proposed MixerNet on CSI recovery accuracy compared with EVCsiNet and Transformer.

3.3.3.4 Complexity Analysis

The model complexity can be described by the trainable parameters and Floating Point Operations (FLOPs). Therefore, the trainable parameters and FLOPs of different models are listed in Table 3.2. Obviously, we can see that our proposed MixerNet has the least trainable parameters and FLOPs. Taking $Q = 120$ as an example, the trainable parameters and FLOPs of MixerNet reduce about 73.3% and 94.1% compared with EVCsiNet and Transformer, respectively. This is due to the large number of convolution layers and attention layers in EVCsiNet and Transformer, respectively, while MixerNet is made up of some simple fully connected layers. In summary, our proposed MixerNet has the lowest computation complexity while ensuring high CSI recovery accuracy.

3.4 Adaptive ABLNet for CSI Feedback

The adaptive CSI feedback model based on BiLSTM structure, named ABLNet, is first introduced in this section. The role of ABLNet includes adapting to different input lengths of CSI eigenvectors and ensuring the recovery accuracy. Then, the FBCU module is designed to make the feedback bit number of ABLNet model more adjustable. Based on which, the BNA algorithm is introduced to achieve the target SGCS for every input CSI by adjusting the number of feedback bits.

3.4.1 Architecture Design of ABLNet

By fully considering the characteristics of input CSI eigenvector and the aim of dealing with different subband numbers, the designed framework of ABLNet is illustrated in Fig. 3.7, which contains encoder and decoder that are described as follows.

Encoder The input CSI eigenvector sequentially passes through four types of blocks, and the specific design of each block is described as follows.

1. *Feature extraction block*: The block consists of two BiLSTM layers, which is responsible for extracting the feature of input CSI eigenvector. The detailed BiLSTM structure is shown in Fig. 3.8, where all subband eigenvectors $\tilde{\mathbf{w}}_1^T, \tilde{\mathbf{w}}_2^T, \cdots, \tilde{\mathbf{w}}_K^T$ are regarded as a sequence. Each subband eigenvector is input to an LSTM cell and is extracted feature vector by using features of preceding and following subband eigenvectors. Then, the K output feature vectors $\mathbf{y}_1, \mathbf{y}_2, \cdots, \mathbf{y}_K$ of $\tilde{\mathbf{w}}_1^T, \tilde{\mathbf{w}}_2^T, \cdots, \tilde{\mathbf{w}}_K^T$ are merged to be transmitted to the

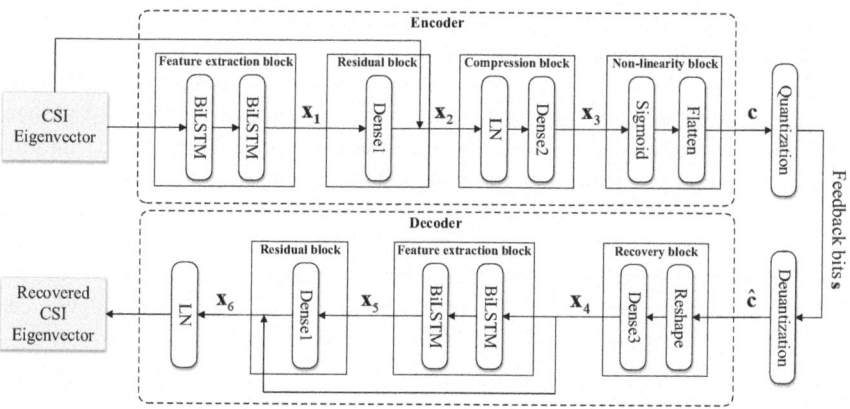

Fig. 3.7 Proposed ABLNet architecture

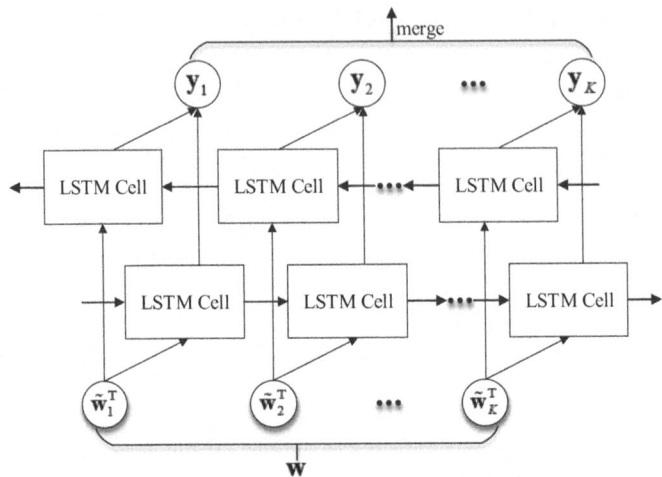

Fig. 3.8 BiLSTM structure

next BiLSTM layer. After passing through this block, the output feature vector can be expressed as $x_1 = \text{BiLSTM}(\text{BiLSTM}(w))$.

2. *Residual block*: The residual block has a fully connected layer, i.e., Dense1, and transforms the output of feature extraction block into Dense1 (x_1) for further processing. Moreover, the residual structure [38] is introduced to accelerate model convergence and improve the model performance. The original input eigenvector w is maintained via the shortcut branch, and so the final output feature vector of the residual block can be written as $x_2 = \text{Dense1}(x_1) + w$.

3. *Compression block*: Data compression is achieved in this block. Firstly, Layer Normalization (LN) operation is carried out to unify the variation range of the extracted feature vector x_2 and help network training. Then, a fully connected layer, Dense2, is employed to compress the output LN (x_2), and therefore, the output of whole compression block can be expressed as $x_3 = \text{Dense2}(\text{LN}(x_2))$.

4. *Nonlinearity block*: In order to utilize the nonlinearity to improve the expressiveness of the model, the sigmoid function is first adopted in the block. And then, the flatten operation is employed to transform its output Sigmoid (x_3) into the codeword vector c, which can be expressed as Flatten (Sigmoid (x_3)).

Finally, the obtained codeword vector c is sent to the quantization layer and is quantized into the feedback bitstream s. In this chapter, the UQ method is adopted to quantize the compressed codeword. Moreover, some other quantization methods introduced in [37], such as nonuniform quantization and VQ, can also be used to improve the CSI feedback performance.

Decoder At the gNB, the dequantization layer is firstly used to transform the received bitstream s into the float codeword vector $\hat{c}$, which is the inverse operation of the quantization layer at the UE. Then, the codeword vector $\hat{c}$ passes through the recovery block, feature extraction block, and residual block in sequence.

1. *Recovery block*: The input codeword vector $\hat{\mathbf{c}}$ is reshaped and then passes through the fully connected layer, i.e., Dense3, and the output vector can be written as $\mathbf{x}_4 = \text{Dense3}\left(\text{Reshape}\left(\hat{\mathbf{c}}\right)\right)$.

   The output vector $\mathbf{x}_4$ has the same size as the original input eigenvector $\mathbf{w}$, while it does not represent the final recovered CSI eigenvector and further processing is needed.

2. *Feature extraction block and residual block*: Similar to the encoder, the feature extraction block and residual block are used to extract the data feature. The input vector $\mathbf{x}_4$ passes through the feature extraction block, and the obtained output feature can be expressed as $\mathbf{x}_5 = \text{BiLSTM}\left(\text{BiLSTM}\left(\mathbf{x}_4\right)\right)$.

   Then, the feature $\mathbf{x}_5$ is fed into the residual block, and also the residual structure is adopted to obtain the output as $\mathbf{x}_6 = \text{Dense1}\left(\mathbf{x}_5\right) + \mathbf{x}_4$.

   After that, the LN operation is adopted to get the final recovered CSI eigenvector $\hat{\mathbf{w}}$, which can be written as $\hat{\mathbf{w}} = \text{LN}\left(\mathbf{x}_6\right)$.

3.4.2   Adaptive Method for CSI Feedback

3.4.2.1   ABLNet for Adjustable Subband Number

1. *Deal with different subband numbers*: Generally, DL-based models have the fixed-length input interface. As a result, in order to satisfy the fixed input length of the proposed ABLNet, the input CSI eigenvectors with a different number of subbands need to be unified to the maximum number of subbands through the approach of padding 0. Meanwhile, the BiLSTM layer only processes the nonzero input part during the feature extraction, and the zero-padding part is automatically ignored. Therefore, on the one hand, every subband eigenvector of each kind of subband number can be further compressed and quantified through the ABLNet after padding operation; on the other hand, the length Q of feedback bitstreams is in proportion to the number K of CSI eigenvector subbands.

   In this way, when the ABLNet model receives CSI eigenvectors with different subband numbers in different scenarios, the feedback overhead can also change with the subband number. As shown in Fig. 3.9, CSI eigenvectors with subband number $K_1, K_2, \cdots, K_M$ are input to the model, and after compression and quantization, the lengths $Q_1, Q_2, \cdots, Q_M$ of feedback bits $\mathbf{s}_1, \mathbf{s}_2, \cdots, \mathbf{s}_M$ are proportional with the corresponding subband number of CSI eigenvectors, i.e.,

$$\frac{Q_1}{K_1} = \frac{Q_2}{K_2} = \cdots = \frac{Q_M}{K_M}. \tag{3.17}$$

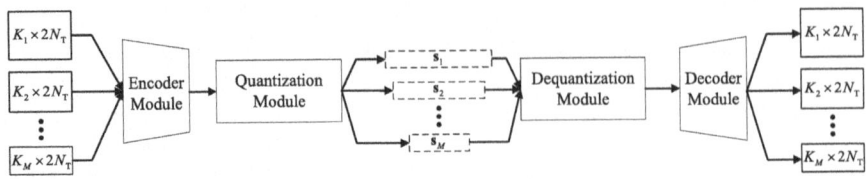

Fig. 3.9 CSI feedback with different lengths of subband eigenvectors

2. *Loss function*: As mentioned in Sect. 3.2, SGCS is used as the loss function of the proposed ABLNet model, which can be formulated as

$$L_1 = \sum_{i=1}^{M} \sum_{k=1}^{K_i} \frac{\mu_i}{K_i} \left(\frac{\| \mathbf{w}_k^H \hat{\mathbf{w}}_k \|}{\| \mathbf{w}_k \| \| \hat{\mathbf{w}}_k \|} \right)^2, \tag{3.18}$$

where $K_i \in \{K_1, K_2, \cdots, K_M\}$ is the number of CSI eigenvector subbands and μ_i is the weight coefficient of SGCS of CSI eigenvector with subband number K_i. In this chapter, when CSI eigenvectors with multiple subband numbers are trained at the same time, equal weight coefficient $\mu_i = 1/M$ is assumed for simplification. However, when the ABLNet is trained for a fixed subband number K_j, the loss function can be set as $\mu_i = 1 \, (i = j)$ and $\mu_i = 0 \, (i \neq j)$.

3.4.2.2 FBCU for Adjustable Feedback Bit Number

Since the encoder output length m of the codeword vector $\mathbf{c}$ and the number q of quantization bit for each float number in the output codeword vector $\mathbf{c}$ are predetermined, the CSI feedback bitstream $\mathbf{s}$ obtained by the UE has a constant length Q, which can be calculated as

$$Q(m) = m \times q. \tag{3.19}$$

In the actual communication systems, different scenarios or applications may require changeable feedback overhead, even for the same length of input eigenvector. And every subband CSI eigenvector can report different lengths of feedback information. Therefore, as shown in Fig. 3.10, a CSI feedback scheme with FBCU that adapts to different numbers of feedback bits is proposed to improve the generalization and practicability of the CSI feedback model.

1. *FBCU principle*: The FBCU is applied for CSI feedback model to change the length of the compressed vector output by the encoder and further change the number of feedback bits. Assuming the output codeword vector of the encoder is $\mathbf{c} = [c_1, c_2, \cdots, c_m]$ in Fig. 3.10, the FBCU can maintain the codeword vector with length n, i.e., $\bar{\mathbf{c}} = [c_1, c_2, \cdots, c_n]$, according to the feedback overhead

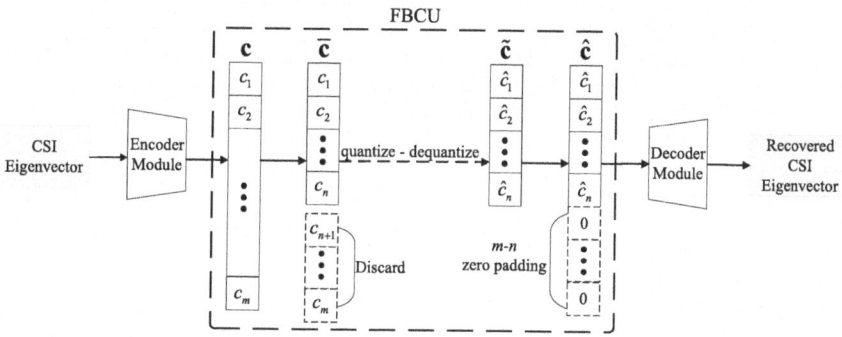

Fig. 3.10 CSI feedback scheme with FBCU

requirements, and directly discard the rest part of the codeword vector, i.e., $[c_{n+1}, c_{n+2}, \cdots, c_m]$. Then, the remaining codeword vector $\bar{\mathbf{c}}$ is quantized and fed back to the gNB through the feedback link. Due to the fact that the length n is variable, the length Q of feedback bitstream $\mathbf{s}$ obtained by quantization also changes with n, which can be written as

$$Q(n) = n \times q. \tag{3.20}$$

At the gNB, after passing the dequantization layer, the dequantized codeword vector $\tilde{\mathbf{c}} = [\hat{c}_1, \hat{c}_2, \cdots, \hat{c}_n]$ is obtained. Similar to the input of the encoder, the input dimension of the dequantized codeword vector to the decoder should also be fixed, and the decoder cannot directly deal with length-variable codeword vector. Therefore, the FBCU should unify the dequantized codeword vector $\tilde{\mathbf{c}}$ with different lengths n into the codeword vector with the maximum length m by the approach of padding zero, i.e., $\hat{\mathbf{c}} = [\hat{c}_1, \hat{c}_2, \cdots, \hat{c}_n, 0, 0 \cdots, 0]$. Finally, the decoder reconstructs the CSI eigenvector $\hat{\mathbf{w}}$ from the zero-padded codeword vector $\hat{\mathbf{c}}$.

2. *ABLNet with FBCU*: Combining FBCU and ABLNet can realize changeable input length of CSI eigenvector and feedback bit number by only one pair of encoder and decoder. Therefore, the ABLNet with FBCU can improve the generalization and practicability of CSI feedback model.

To train the proposed ABLNet with FBCU, the length set $\mathbf{N}_i$ ($1 \le i \le M$) of codeword vector $\bar{\mathbf{c}}$ for the CSI eigenvectors with subband number K_i should be first determined. Then, during the training process, a specific length $n_i \in \mathbf{N}_i$ is randomly selected in each training epoch. Once the length n_i is fixed, the FBCU discards the last $m - n_i$ elements of the codeword vector $\mathbf{c}$ and pad zeros to the dequantized and truncated codeword vector $\tilde{\mathbf{c}}$ to guarantee the same input length m of the codeword vector $\hat{\mathbf{c}}$ for the decoder. In this way, the SGCS can still be used as the loss function, which is the same as (3.18).

Moreover, since the FBCU only discards part of the compressed codeword vector $\mathbf{c}$ to change the length of feedback bitstream, it can also be applied to any

---

Algorithm 1 BNA algorithm

Input: Initial feedback bit number Q_i and corresponding SGCS performance ρ_i; Target SGCS
performance ρ_t; Supported minimum feedback bit number Q_{min} and maximum feedback
bit number Q_{max}; Tolerance Error $\varepsilon = 0.01$

Output: Adjusted feedback bit number Q_a and corresponding SGCS performance ρ_a

if $|\rho_i - \rho_t| \le \varepsilon$ **then**
| $[Q_a, \rho_a] = [Q_i, \rho_i]$;
else if $\rho_i - \rho_t > \varepsilon$ **then**
| $[Q_a, \rho_a] = Binary\_Search(Q_{min}, Q_i)$;
else if $\rho_i - \rho_t < -\varepsilon$ **then**
| $[Q_a, \rho_a] = Binary\_Search(Q_i, Q_{max})$;
end
return Q_a, ρ_a;

Function $Binary\_Search(Q_l, Q_r)$:
   | **while** $Q_l \le Q_r$ **do**
      | $Q_{mid} = (Q_l + Q_r)/2$;
      | Calculate the SGCS ρ_{mid} as (3.9) with Q_{mid};
      | **if** $|\rho_{mid} - \rho_t| \le \varepsilon$ **then**
        | **return** Q_{mid}, ρ_{mid};
      | **else if** $\rho_{mid} - \rho_t < -\varepsilon$ **then**
        | $Q_l = Q_{mid} + 1$;
      | **else**
        | $Q_r = Q_{mid} - 1$;
      | **end**
   | **end**
   | **return** Q_{mid}, ρ_{mid};
end

DL-based CSI feedback model mentioned above [14–21] to realize the adjustable
number of feedback bits.

3.4.2.3   BNA Algorithm for Target SGCS

1. *Purpose of BNA algorithm*: Generally, when the CSI feedback performance
   approaches one target SGCS ρ_t, further improving the SGCS will increase
   little communication performance and meanwhile consume a lot of feedback
   overhead. On the other hand, for the varying input CSI eigenvectors, the SGCS
   performance is different with a fixed number of feedback bits. Therefore, by
   utilizing the adjustable characteristic of feedback bit numbers of ABLNet with
   FBCU, a BNA algorithm is developed to increase or decrease the number Q of
   feedback bits to make the SGCS performance ρ stabilize at a reasonable target
   SGCS ρ_t for every feedback eigenvector.
2. *Description of BNA algorithm*: Assume that the minimum and maximum numbers of feedback bits supported by ABLNet with FBCU are Q_{min} and Q_{max},
   respectively; given the initial feedback bit number Q_i, the corresponding SGCS
   performance ρ_i, the target SGCS ρ_t, and the tolerance error $\varepsilon = 0.01$, the BNA
   algorithm is shown in Algorithm 1.

In Algorithm 1, according to the relation of $|\rho_i - \rho_t|$ and ε, the *Binary_Search* method is taken to find the appropriate feedback bit number Q_a, making the corresponding feedback performance ρ_a within the tolerance error ε of ρ_t. Because the SGCS performance ρ increases with the feedback bit number Q, taking the binary search strategy can improve the adjustment efficiency. Finally, both Q_a and ρ_a are the output of the BNA algorithm as the final adjusted feedback bit number and SGCS performance, respectively.

Moreover, if the ρ_a has reached the error ε range of ρ_t, but the adjusted feedback bit number Q_a is lower than Q_{min} or higher than Q_{max}, the BNA algorithm takes the Q_{min} or Q_{max} and the corresponding feedback performance as the final output.

3.4.3 Performance Evaluation

In this section, the dataset, the relevant training setup, and the structure of networks are first introduced in detail. Then, a series of experiments are implemented to evaluate the performance of the proposed ABLNet for the CSI eigenvectors with different numbers of subbands. Furthermore, the performance of ABLNet with FBCU is discussed for length-variable feedback bitstreams. After that, the adjustment effect of the BNA algorithm is illustrated. Finally, the link-level Block Error Rate (BLER) performance with different feedback schemes is analyzed.

3.4.3.1 Simulation Configuration

The CDL channel model defined in the 3GPP is adapted to generate data samples, and two types of channels with delay spread 30ns are taken into consideration: CDLA and CDLC. Each channel contains three kinds of subband numbers for CSI eigenvector, which are $K = 3$ subbands, $K = 6$ subbands, and $K = 12$ subbands. Moreover, a typical setup that $N_T = 32$ transmit antennas at the gNB and $N_R = 4$ receive antennas at UE is adopted in the MIMO-OFDM system. Therefore, the size of CSI eigenvector is $K \times 2N_T = K \times 64\,(K = 3, 6, 12)$.

At the same time, the sizes of training and testing datasets are 50,000 and 1000, respectively, for CSI eigenvector with any subband number. For the training phase, we set the numbers of training epochs and batch size as 500 and 128, respectively. The initial learning rate is 0.0005, which will be adaptively adjusted with the training process. The default Adam optimizer is employed. Because the proposed FBCU is a plug-in module and compatible with the proposed network, so the training setup is the same as the original setup mentioned above when training models with FBCU. Additionally, we fix the quantization bit number $q = 2$.

The detailed model structure of ABLNet is indicated in Table 3.3. Because the input layer of the model is fixed, CSI eigenvectors with different subband numbers

Table 3.3 ABLNet model structure

Model	Structure
Encoder	Input layer, input $= (K_{\max}, 64)$
	BiLSTM, input $= (K_{\max}, 64)$, output $= (K_{\max}, 256)$
	BiLSTM, input $= (K_{\max}, 256)$, output $= (K_{\max}, 1024)$
	Dense1, input $= (K_{\max}, 1024)$, output $= (K_{\max}, 64)$
	Dense2, input $= (K_{\max}, 64)$, output $= (K_{\max}, 5)$
	Quantization layer, input $= K \times 5$, output $= K \times 10$
Decoder	Dequantization layer, input $= K \times 10$, output $= K \times 5$
	Dense3, input $= (K_{\max}, 5)$, output $= (K_{\max}, 64)$
	BiLSTM, input $= (K_{\max}, 64)$, output $= (K_{\max}, 256)$
	BiLSTM, input $= (K_{\max}, 256)$, output $= (K_{\max}, 1024)$
	Dense1, input $= (K_{\max}, 1024)$, output $= (K_{\max}, 64)$
	Output layer, output $= (K, 64)$

should be unified into the same dimension $K_{\max} \times 64$ by padding zero according to the maximum subband number $K_{\max}$.

Furthermore, to verify the SGCS performance of the ABLNet and FBCU with different subband numbers of input CSI eigenvectors, the designed model structure is utilized to train CSI eigenvectors with a fixed subband number, denoted as BiLSTMNet. And the obtained SGCS performance of BiLSTMNet is used as the benchmark.

3.4.3.2   Performance of Proposed ABLNet

The number of feedback bits is assumed to be $Q = 120$ bits for CSI eigenvector with $K = 12$ subbands, and then the corresponding numbers of CSI feedback bits for $K = 3$ subbands eigenvector and $K = 6$ subbands eigenvector are $Q = 30$ bits and $Q = 60$ bits, respectively.

To verify the feedback performance of the ABLNet model, different comparison models, such as EVCsiNet and Transformer [20, 21], for CSI eigenvector with different subband numbers are trained under both CDLA and CDLC channels, as illustrated in Fig. 3.11. Concretely, the adaptive input scheme in [39] is adopted to train both EVCsiNet and Transformer. In Fig. 3.11, it can be seen that the feedback performance of ABLNet is better than that of BiLSTMNet and EVCsiNet under $K = 3$ subbands and $K = 6$ subbands and similar to the Transformer under $K = 12$ subbands for CDLA channel. Similarly, the feedback performance of ABLNet for CDLC channel outperforms those of other models under any kind of subband number.

Moreover, the SGCS gain of low subband number is more obvious. Because compared with the BiLSTMNet trained by a fixed subband number alone, the ABLNet trained with multiple subband numbers can learn more data features of

Fig. 3.11 Feedback performance with different subband numbers

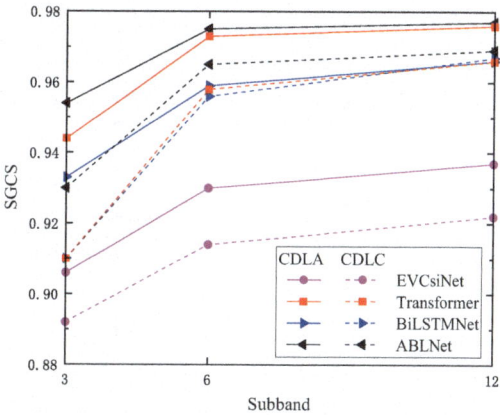

Table 3.4 Model complexity comparison

Feedback scheme	Trainable parameters ($\times 10^6$)	FLOPs ($\times 10^7$)
EVCsiNet	6.98	1.41
Transformer	23.95	51.29
ABLNet	14.75	32.40

other subband. Additionally, with the increase of subband number for each curve in Fig. 3.11, the feedback performance also increases gradually, because more subbands make CSI eigenvector contain more features that can be extracted during training.

Additionally, as listed in Table 3.4, the model complexities of different feedback schemes are analyzed in terms of trainable parameters and FLOPs. We can see that the proposed ABLNet has fewer trainable parameters and FLOPs than Transformer but has better feedback performance. Although EVCsiNet has the fewest trainable parameters and FLOPs, its performance is worst among these feedback schemes.

Therefore, simulation results show that the proposed ABLNet can adapt to the CSI eigenvector with different subband numbers and has a low model complexity while high CSI recovery accuracy; moreover, the feedback bit number can be proportionally varied with the number of subbands.

3.4.3.3 Performance of ABLNet with FBCU

The performance of ABLNet with FBCU is tested under the feedback bit number $Q = 20$ to 30 for $K = 3$ subbands, $Q = 40$ to 60 for $K = 6$ subbands, and $Q = 80$ to 120 for $K = 12$ subbands, respectively.

Figure 3.12 illustrates the SGCS curves under both the ABLNet with FBCU and BiLSTMNet with FBCU. Moreover, the SGCS performance of EVCsiNet and Transformer with single encoder and multiple decoders is taken as the benchmark [20, 21, 29]. To save the training cost, three decoders for EVCsiNet and Transformer

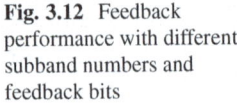

Fig. 3.12 Feedback
performance with different
subband numbers and
feedback bits

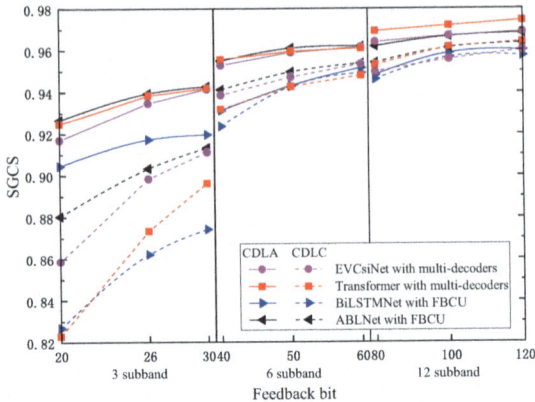

are employed, as well as $Q = \{20, 26, 30\}$ for $K = 3$ subbands, $Q = \{40, 50, 60\}$
for $K = 6$ subbands, and $Q = \{80, 100, 120\}$ for $K = 12$ subbands, respectively.
Whether for CDLA or CDLC channel, it is obvious that the proposed ABLNet
trained with FBCU outperforms other training approaches under $K = 3$ subbands
and $K = 6$ subbands, while for $K = 12$ subbands, the SGCS performance
of ABLNet with FBCU is slightly less than that of EVCsiNet and Transformer
with multi-decoders. The reason has been mentioned above that data features of
eigenvectors with small subband number have a negative impact on the eigenvectors
with large subband number during the training process.

At the same time, the SGCS performance is gradually improved with the increase
of feedback bit number for each subband number for each training approach,
because the longer the feedback bitstream is, the more feature information it
contains, and this will help the decoder to reconstruct the CSI eigenvectors.

In summary, simulation results demonstrate that the FBCU module is compatible
with the proposed ABLNet model and makes the model adapt to different subband
numbers of CSI eigenvectors and different feedback bit numbers at the same time.

3.4.3.4   Performance of BNA Algorithm

To verify the adjustment effect of BNA algorithm, the ABLNet with FBCU is first
trained under $Q = 20$ to 120 for $K = 12$ subbands. Then, $Q = 49$ of Enhanced
Type II (eTypeII) method is taken as the benchmark, and the corresponding target
SGCS performances are $\rho_t = 0.904$ and $\rho_t = 0.840$ for CDLA and CDLC channel,
respectively [20]. Additionally, the initial feedback bit number $Q_i = 48$ is chosen
for ABLNet with FBCU. The other subband numbers have the similar conclusions
and therefore are ignored here.

The Cumulative Distribution Function (CDF) is taken to illustrate the SGCS
performance changes of input CSI before and after the adaptive adjustment, as
ρ_i and ρ_a illustrated in both Figs. 3.13 and 3.14. From both figures, it can be
obviously noticed that the CDF of ρ_a is higher than that of ρ_i within the tolerance

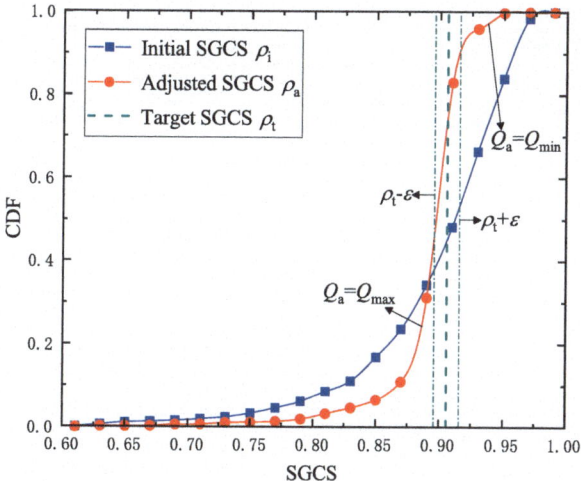

Fig. 3.13 Adjustment effect of BNA algorithm for CDLA

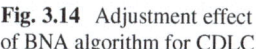

Fig. 3.14 Adjustment effect of BNA algorithm for CDLC

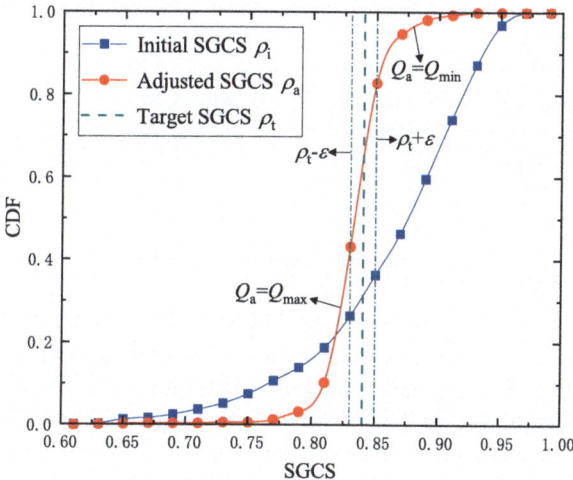

error margin of ρ_t, meaning that the SGCS performances of most input CSIs are effectively adjusted close to the target performance ρ_t through the BNA algorithm. Specifically, when the SGCS ρ_a reaches $\rho_t + \varepsilon$, the CDF improves about 40% and 50% compared with ρ_i for CDLA and CDLC, respectively. Moreover, the feedback bit number of input CSI eigenvectors with SGCS lower than $\rho_t - \varepsilon$ has been adjusted as $Q_a = Q_{max}$, which has improved the SGCS performance of every input CSI as much as possible; and for the input CSI eigenvectors with SGCS higher than $\rho_t + \varepsilon$, the feedback bit number has been adjusted to $Q_a = Q_{min}$ for saving the average feedback overhead. Additionally, the average value of feedback bit number Q_a is

Fig. 3.15 Adjustment effect of BNA algorithm for CDLC

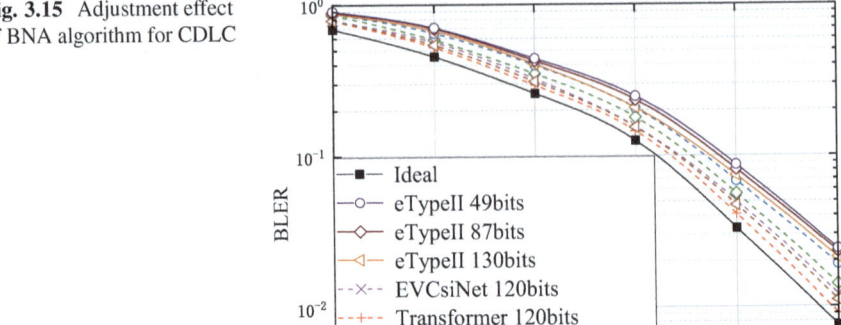

also calculated, which is 44 bits for both two channels and save 4 bits compared with Q_i.

Therefore, the results demonstrate that the BNA algorithm can adaptively adjust the feedback bit number of every input CSI to make the SGCS performance close to the target SGCS performance ρ_t and can also save the average feedback bit number Q_a after adjustment.

3.4.3.5 Performance of Link-Level BLER

As shown in Fig. 3.15, the link-level BLER for different CSI feedback schemes is depicted under CDLA channel. The ideal condition means that the gNB knows the CSI eigenvector, and therefore, the SGCS $\rho = 1$. Meanwhile, the BLER performance of $Q = \{40, 80, 120\}$ for ABLNet with FBCU is compared to those of EVCsiNet and Transformer at $Q = 120$ bits. Moreover, for the traditional CSI feedback scheme, i.e., eTypeII, the number $Q = \{49, 87, 130\}$ of feedback bits is close to those of DL-based CSI feedback schemes and therefore is adopted here for comparison.

It can be seen that the proposed ABLNet with FBCU achieves about BLER $= 10^{-2}$ when $Q = 120$ and SNR $= 5$, which significantly outperforms eTypeII with $Q = 130$ bits; the proposed ABLNet with FBCU and $Q = 40$ bits can achieve similar BLER performance to eTypeII with $Q = 130$ bits. As the feedback bit number Q increases for both the ABLNet with FBCU and eTypeII, the BLER performance decreases gradually, because the gNB can select a more appropriate beamforming vector with larger SGCS performance for each input CSI eigenvector, which is more beneficial to the average BLER reduction. Moreover, it can be noticed that three different DL-based CSI feedback models have similar BLER performance with $Q = 120$ feedback bits.

Therefore, the results show the superiority of the proposed ABLNet with FBCU on the BLER performance and feedback overhead compared with conventional eTypeII method. Additionally, the proposed ABLNet with FBCU can obtain the similar BLER performance of other DL-based CSI feedback schemes with the capability of adaptive input length of CSI eigenvectors and output number of feedback bits.

3.5 Summary

In order to relieve the CSI feedback overhead of massive MIMO-OFDM systems, the MixerNet model is first designed based on MLP-Mixer, where both the encoder and decoder are studied in detail. On the other hand, two kinds of quantization methods, i.e., UQ and VQ, are discussed for the output of encoder in the proposed MixerNet model. Then, an adaptive CSI feedback model ABLNet is proposed, and an FBCU is designed to process different lengths of input CSI eigenvectors and the numbers of feedback bits. Subsequently, a BNA algorithm is developed to achieve a target SGCS for every input CSI by adjusting feedback bit number flexibly. Experiments reveal that the proposed MixerNet not only improves the accuracy of CSI recovery, but also significantly reduces the model complexity, including both the trainable parameters and FLOPs. And the VQ method has a better performance on CSI recovery than UQ method. Moreover, the proposed ABLNet has better feedback performance with adapting different lengths of input CSI. Meanwhile, ABLNet with FBCU can improve the model robustness and maintain the SGCS performance. Finally, the designed BNA algorithm can effectively stabilize the SGCS performance for every input CSI with fewer number of feedback bits.

References

1. Pereira de Figueiredo, F.A.: An overview of massive MIMO for 5G and 6G. IEEE Latin Am. Trans. **20**(6), 931–940 (2022)
2. Liyanaarachchi, S.D., Riihonen, T., Barneto, C., Valkama, M.: Optimized waveforms for 5G–6G communication with sensing: theory, simulations and experiments. IEEE Trans. Wirel. Commun. **20**(12), 8301–8315 (2021)
3. Shi, Y., Badi, M., Rajan, D., Camp, J.: Channel reciprocity analysis and feedback mechanism design for mobile beamforming systems. IEEE Trans. Veh. Technol. **70**(6), 6029–6043 (2021)
4. Eltayeb, M.E., Al-Naffouri, T.Y., Bahrami, H.R.: Compressive sensing for feedback reduction in MIMO broadcast channels. IEEE Trans. Commun. **62**(9), 3209–3222 (2014)
5. Huang, X., Wu, J., Wen, Y., Hu, F., Wang, Y., Jiang, T.: Rate-adaptive feedback with Bayesian compressive sensing in multiuser MIMO beamforming systems. IEEE Trans. Wirel. Commun. **15**(7), 4839–4851 (2016)

6. Wang, Y., Tan, H., Wu, Y., Peng, J.: Hybrid electric vehicle energy management with computer vision and deep reinforcement learning. IEEE Trans. Ind. Inform. **17**(6), 3857–3868 (2021)
7. Yang, A., Zhi, J., Yang, K., Wang, J., Xue, T.: Computer vision technology based on sensor data and hybrid deep learning for security detection of blast furnace bearing. IEEE Sensors J. **21**(22), 24982–24992 (2021)
8. Otter, D.W., Medina, J.R., Kalita, J.K.: A survey of the usages of deep learning for natural language processing. IEEE Trans. Neural Netw. Learn. Syst. **32**(2), 604–624 (2021)
9. Liu, T., Lei, L., Zheng, K., Shen, X.: Multitimescale control and communications with deep reinforcement learning—Part I: communication-aware vehicle control. IEEE Internet Things J. **11**(9), 15386–15401 (2024)
10. Lei, L., Liu, T., Zheng, K., Shen, X.: Multitimescale control and communications with deep reinforcement learning—Part II: control-aware radio resource allocation. IEEE Internet Things J. **11**(9), 15475–15489 (2024)
11. Zheng, K., Yang, H., Ying, Z., Wang, P., Hanzo, L.: Vision-assisted millimeter-wave beam management for next-generation wireless systems: concepts, solutions, and open challenges. IEEE Veh. Technol. Mag. **18**(3), 58–68 (2023)
12. Cheng, X., Liu, D., Wang, C., Yan, S., Zhu, Z.: Deep learning-based channel estimation and equalization scheme for FBMC/OQAM systems. IEEE Wirel. Commun. Lett. **8**(3), 881–884 (2019)
13. Xiang, B., Hu, D., Wu, J.: Deep learning-based downlink channel estimation for FDD massive MIMO systems. IEEE Wirel. Commun. Lett. **12**(4), 699–702 (2023)
14. Wen, C., Shi, W., Jin, S.: Deep learning for massive MIMO CSI feedback. IEEE Wirel. Commun. Lett. **5**(11), 748–751 (2018)
15. Cai, Q., Dong, C., Niu, K.: Attention model for massive MIMO CSI compression feedback and recovery. In: IEEE Wireless Communications and Networking Conference, pp. 1–5 (2019)
16. Liu, Z., Zhang, L., Ding, Z.: Exploiting bi-directional channel reciprocity in deep learning for low rate massive MIMO CSI feedback. IEEE Wirel. Commun. Lett. **8**(3), 889–892 (2019)
17. He, Z., Zhao, L., Luo, X., Cheng, B.: Deep CSI feedback for FDD MIMO systems. In: Communications and Networking, pp. 366–376 (2022)
18. Liu, Z., Rosario, M., Liang, X., Zhang, L., Ding, Z.: Spherical normalization for learned compressive feedback in massive MIMO CSI acquisition. In: IEEE International Conference on Communications Workshops, pp. 1–6 (2020)
19. Lu, C., Xu, W., Jin, S., Wang, K.: Bit-level optimized neural network for multi-antenna channel quantization. IEEE Wirel. Commun. Lett. **9**(1), 87–90 (2020)
20. Liu, W., Tian, W., Xiao, H., Jin, S., Liu, X., Shen, J.: EVCsiNet: Eigenvector-based CSI feedback under 3GPP link-level channels. IEEE Wirel. Commun. Lett. **10**(23), 2688–2692 (2021)
21. Vaswani, A., Shazeer, N., Parmar, N., et al.: Attention is all you need. In: 31st Conference on Neural Information Processing Systems (NIPS) (2017)
22. Dehdashtian, S., Hashemi, M., Salehkaleybar, S.: Deep-learning-based blind recognition of channel code parameters over candidate sets under AWGN and multi-path fading conditions. IEEE Wirel. Commun. Lett. **10**(5), 1041–1045 (2021)
23. Xie, Y., Teh, K.C., Kot, A.C.: Deep learning-based joint detection for OFDM-NOMA scheme. IEEE Commun. Lett. **25**(8), 2609–2613 (2021)
24. Ji, D., Park, J., Cho, D.: ConvAE: a new channel autoencoder based on convolutional layers and residual connections. IEEE Commun. Lett. **23**(10), 1769–1772 (2019)
25. 3GPP: 3rd generation partnership project; technical specification group radio access network; NR; physical layer procedures for data (release 16). Sophia Antipolis, France, TS 38.214 V16.1.0, 2020 (2020)
26. Yang, Q., Mashhadi, M.B., Gündüz, D.: Deep convolutional compression for massive MIMO CSI feedback. In: IEEE 29th International Workshop on Machine Learning for Signal Processing (MLSP), pp. 1–6 (2019)
27. Mashhadi, M.B., Yang, Q., Gündüz, D.: Distributed deep convolutional compression for massive MIMO CSI feedback. IEEE Trans. Wirel. Commun. **20**(4), 2621–2633 (2021)

28. Cao, Z., Shih, W.T., Guo, J., Wen, C., Jin, S.: Lightweight convolutional neural networks for CSI feedback in massive MIMO. IEEE Commun. Lett. **25**(8), 2624–2628 (2021)
29. Guo, J., Wen, C., Jin, S., Li, G.: Convolutional neural network-based multiple-rate compressive sensing for massive MIMO CSI feedback: Design, simulation, and analysis. IEEE Trans. Wirel. Commun. **19**(4), 2827–2840 (2020)
30. Wang, Y., Zhang, Y., Sun, J., Gui, G., Adachi, F.: A novel compression CSI feedback based on deep learning for FDD massive MIMO systems. In: IEEE Wireless Communications and Networking Conference (WCNC), pp. 1–5 (2021)
31. Liang, X., Chang, H., Li, H., Gu, X., Zhang, L.: Changeable rate and novel quantization for CSI feedback based on deep learning. IEEE Trans. Wirel. Commun. **21**(12), 10100–10114 (2021)
32. Jo, S., So, J.: Adaptive lightweight CNN-based CSI feedback for massive MIMO systems. IEEE Wirel. Commun. Lett. **10**(12), 2776–2780 (2021)
33. Lu, Z., Zhang, L., Ding, Z.: An efficient deep learning framework for low rate massive MIMO CSI reporting. IEEE Trans. Commun. **68**(8), 4761–4772 (2020)
34. Cui, T., Tellambura, C.: Semiblind channel estimation and data detection for OFDM systems with optimal pilot design. IEEE Trans. Commun. **55**(5), 1053–1062 (2007)
35. Tolstikhin, I., Houlsby, N., Kolesnikov, A., et al.: MLP-mixer: an all-MLP architecture for vision (2021). https://arxiv.org/abs/2105.01601
36. Oord, A., Vinyals, O., Kavukcuoglu, K.: Neural discrete representation learning (2017)
37. Zhou, B., Ma, S., Yang, G.: Transformer-based CSI feedback with hybrid learnable non-uniform quantization for massive MIMO systems. In: 32nd Wireless and Optical Communications Conference (WOCC), pp. 1–5 (2023)
38. Liu, W., Wu, G., Ren, F., Kang, X.: DFF-ResNet: An insect pest recognition model based on residual networks. Big Data Mining Anal. **3**(4), 300–310 (2020)
39. R1-2211189: Evaluation methodology and results on AI/ML for CSI feedback enhancement. RAN1#111, CATT (2022)

Chapter 4
Intelligent Precoding Technology for TDD Systems

Abstract It is hard for traditional precoding algorithms to balance the high performance and computational complexity. Therefore, a low-complexity MLP-mixer-based precoding network (MMPNet) is proposed with unsupervised training in this chapter. The feasibility of using the compressed channel correlation matrix as input of the DL network is first proved. Subsequently, a low-complexity feature extraction module is designed to fit for the input structure, and the recovery module of the WMMSE algorithm is also improved. Finally, the scalability of the proposed model versus the varying numbers of UEs is discussed.

Keywords AI · Precoding · TDD · MLP-Mixer · WMMSE

4.1 Background

MIMO is deployed in Fifth-Generation New Radio (5G-NR) and beyond to handle the large number of UEs and the enormous data traffic [1–3]. The increased number of antennas significantly raises the computational overhead of precoding. Consequently, low-complexity precoding technology is essential for the wireless communication systems for 5G and beyond.

In traditional MIMO systems, many well-known precoding algorithms have been proposed to improve system performance. The Weighted Minimum Mean Square Error (WMMSE) algorithm has achieved outstanding performance using the block coordinate descent method [4–6]. However, the large amount of inversion and iteration results in its high computational complexity. The ZF precoder and the Block Diagonalization (BD) precoder project signals onto the null space of the channels of other UEs [7, 8], which can completely eliminate inter-UE interference. In addition, Multiuser Eigenmode Transmission (MET) precoder, improved from the BD precoder, can serve more UEs through matrix compression with negligible performance loss [9, 10]. The three types of precoders have lower complexity, however, a significant gap compared to WMMSE algorithm. Therefore, it is difficult for traditional precoding to simultaneously offer high performance and low computational complexity.

The powerful feature extraction and data fitting capabilities of DL make it widely used in wireless communication fields, such as channel estimation, CSI feedback, and resource management [11–13]. Numerous DL-based precoding schemes have been proposed to overcome the shortcomings of traditional precoding [14–18]. Inspired by WMMSE algorithm, an iterative unfolding network was proposed, which replaces matrix inversion with the special designed linear operations, significantly reducing its complexity and achieving excellent performance [14]. However, the network requires specially designed backpropagation algorithms, meaning that any modifications for new scenarios necessitate changes to the backpropagation algorithm. Then, based on WMMSE algorithm, [15] designed a CNN that takes the intermediate variables of WMMSE as the output and obtains the precoding through post-processing of the intermediate variables. Moreover, a hybrid network was introduced, consisting of a model-driven subnetwork that simulates Minimum Mean Square Error (MMSE) algorithm and a data-driven fully connected neural subnetwork [16]. The model-driven subnetwork is used to augment the CSI, thereby alleviating the learning burden of the data-driven subnetwork and improving the overall performance of hybrid networks. Nevertheless, the performance of the models proposed in [15] and [16] still performs worse than that of WMMSE algorithm.

Existing research of DL-based precoding primarily utilizes CNNs, resulting in a significant number of redundant parameters, which introduce unnecessary computational overhead and impact model performance [15–18]. Furthermore, the training data is often generated by simplified channel models, deviating from realistic communication scenarios. Finally, most studies employ a combination of supervised and unsupervised learning to enhance model performance. However, it is challenging to obtain the labels for supervised learning due to the high computational cost of optimal or suboptimal precoding labels.

To address the aforementioned issues, a low-complexity MLP-Mixer-based precoding network (MMPNet) with unsupervised training is proposed in this chapter and consists of model-driven preprocessing module, data-driven feature extraction module, and model-driven recovery module [19]. Initially, the preprocessing module compresses the channel matrix into a variant of the channel correlation matrix to remove redundant information; subsequently, the feature extraction module extracts the key feature matrices from the compressed CSI; and the recovery module ultimately calculates the precoding matrix based on the WMMSE algorithm. Simulation results illustrate that the proposed scheme achieves close performance to WMMSE algorithm, while could significantly reduce the computational complexity. Furthermore, it could be adaptable across a range of UE numbers and SNRs.

4.2 System Model and MMPNet Framework

4.2.1 System Model and Problem Formulation

A 5G gNB equipped with N_T transmit antennas serves K UEs equipped with N_R receiving antennas. Assuming that $\mathbf{H}_k \in \mathbb{C}^{N_R \times N_T}$, $\mathbf{V}_k \in \mathbb{C}^{N_T \times r}$, r, and $\mathbf{s}_k \in \mathbb{C}^{r \times 1}$, respectively, denote the channel matrix from the gNB to the kth UE, the downlink precoding matrix for the kth UE, the number of datastreams, and the transmitted data for the kth UE with $\mathbb{E}[\mathbf{s}_k \mathbf{s}_k^H] = \mathbf{I}$, then the signal received by the kth UE can be given by

$$\mathbf{y}_k = \mathbf{H}_k \mathbf{V}_k \mathbf{s}_k + \sum_{i=1, i \neq k}^{K} \mathbf{H}_k \mathbf{V}_i \mathbf{s}_i + \mathbf{n}_k, \tag{4.1}$$

where $\mathbf{n}_k \sim \mathcal{CN}(0, \sigma_k^2)$ denotes a circularly symmetric complex Gaussian random vector with zero mean and variance σ_k^2 at the kth UE.

Based on the transmission model in (4.1), the rate R_k of the kth UE can be written as

$$R_k = \log\det\left[\mathbf{I} + \mathbf{H}_k \mathbf{V}_k \mathbf{V}_k^H \mathbf{H}_k^H \times \left(\sum_{i \neq k} \mathbf{H}_k \mathbf{V}_i \mathbf{V}_i^H \mathbf{H}_k^H + \sigma_k^2 \mathbf{I} \right)^{-1} \right]. \tag{4.2}$$

Usually, the performance of precoding schemes is evaluated by the system sum rate under the constraint of transmit power P_T, which could be formulated by

$$\max_{\{\mathbf{V}_k\}} R = \sum_{k=1}^{K} R_k$$
$$s.t. \quad \sum_{k=1}^{K} \text{tr}(\mathbf{V}_k \mathbf{V}_k^H) \leq P_T. \tag{4.3}$$

However, the formulated problem (4.3) is nonconvex and therefore extremely hard to find the optimal solution. Many traditional methods have been proposed to obtain an effective solution but have failed to achieve high performance within limited computational complexity.

Therefore, by designing a DL-based precoding network model and optimizing the model parameters set θ_F, the main objective of this chapter is to maximize the sum rate under the power constraint P_T, namely,

$$\{\theta_F\} = \arg\max_{\theta_F} R(\mathbf{H}, \mathbf{V}). \tag{4.4}$$

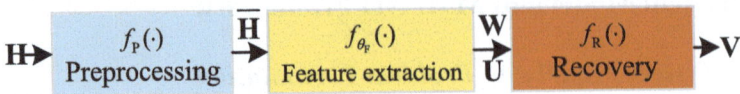

Fig. 4.1 Framework of DL-based precoding model

4.2.2   MMPNet Framework

As shown in Fig. 4.1, the designed MMPNet consists of three parts: data preprocessing module, feature extraction module, and recovery module.

The preprocessing module $f_P(\cdot)$ is model-driven and designed based on expert knowledge, which preserves useful information provided by the CSI. After the preprocessing module, the channel matrix of all UEs $\mathbf{H} \triangleq \left[\mathbf{H}_1^T, \mathbf{H}_2^T, \cdots, \mathbf{H}_K^T\right]^T$ is compressed into an intermediate matrix:

$$\bar{\mathbf{H}} = f_P(\mathbf{H}). \tag{4.5}$$

Then, the feature extraction module $f_{\theta_F}(\cdot)$ with the parameters set θ_F is data-driven, which could learn the mapping from $\bar{\mathbf{H}}$ to the key features. The key features include the receiving matrix $\mathbf{U} \triangleq \mathrm{diag}\left[\mathbf{U}_1, \mathbf{U}_2, \cdots, \mathbf{U}_K\right], \mathbf{U}_k \in \mathbb{C}^{N_R \times r}$ and the weight matrix $\mathbf{W} \triangleq \mathrm{diag}\left[\mathbf{W}_1, \mathbf{W}_2, \cdots, \mathbf{W}_K\right], \mathbf{W}_k \in \mathbb{C}^{r \times r}$, i.e.,

$$(\mathbf{W}, \mathbf{U}) = f_{\theta_F}(\bar{\mathbf{H}}). \tag{4.6}$$

At the end, the model-driven recovery module $f_R(\cdot, \cdot)$ outputs the precoding matrix $\mathbf{V}$ of all UEs:

$$\mathbf{V} = \left[\mathbf{V}_1, \mathbf{V}_2, \cdots, \mathbf{V}_K\right] = f_R(\mathbf{W}, \mathbf{U}). \tag{4.7}$$

4.3   MMPNet Design for Intelligent Precoding

In this section, the preprocessing module, feature extraction module, and precoding recovery module of the proposed MMPNet are first introduced. Then, the unsupervised loss function is designed to improve the training efficiency. Finally, we discussed the generalization ability of the proposed model.

4.3.1   Data Preprocessing

This subsection first introduces the data preprocessing, which will result in information loss, and therefore its feasibility is then discussed.

Fig. 4.2 Preprocessing module

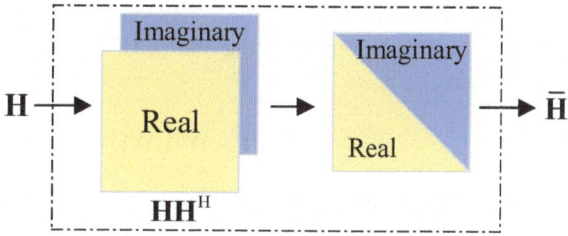

4.3.1.1 Preprocessing Method

As shown in Fig. 4.2, in the preprocessing module, the channel correlation matrix $\mathbf{HH}^H$ is first calculated. Since $\mathbf{HH}^H$ is Hermitian, it can be further lossless compressed using the approach outlined in [15]. We concatenate the lower triangle of its real part with the upper triangle of its imaginary part to obtain the input for the feature extraction module, which is denoted as $\bar{\mathbf{H}}$.

The purpose of data preprocessing is to eliminate redundant information for network inputs and compress input dimensions, thereby facilitating the key feature extraction. Compared with the original channel matrix $\mathbf{H} \in \mathbb{C}^{KN_R \times N_T}$, $\bar{\mathbf{H}} \in \mathbb{C}^{KN_R \times KN_R}$ generally has a smaller size due to $KN_R \leq N_T$ in most cases.

It is worth noting that the channel correlation matrix $\mathbf{HH}^H$ contains less information than $\mathbf{H}$. Next, it is demonstrated that $\mathbf{HH}^H$ contains all requisite information for the key features, including the receiving matrix $\mathbf{U}$ and weight matrix $\mathbf{W}$.

4.3.1.2 Feasibility Proof

In WMMSE algorithm, the receiving matrix $\mathbf{U}$ and weight matrix $\mathbf{W}$ can be initialized as an all-one matrix and an identity matrix, respectively, and the precoding $\mathbf{V}$, new $\mathbf{W}$, and new $\mathbf{U}$ are iteratively updated until the sum rate no longer increases. The process of one iteration is given as follows:

$$\mathbf{V}_k = \left(\beta\mathbf{I} + \mathbf{H}^H\mathbf{UWU}^H\mathbf{H}\right)^{-1}\mathbf{H}_k^H\mathbf{U}_k\mathbf{W}_k, \ \forall k, \tag{4.8}$$

$$\mathbf{U}_k = \left[\frac{\sigma_k^2}{P_T}\sum_{i=1}^{K}\mathrm{Tr}\left(\mathbf{V}_i\mathbf{V}_i^H\right)\mathbf{I} + \sum_{i=1}^{K}\mathbf{H}_k\mathbf{V}_i\mathbf{V}_i^H\mathbf{H}_k^H\right]^{-1}\mathbf{H}_k\mathbf{V}_k, \ \forall k, \tag{4.9}$$

$$\mathbf{W}_k = \left[\mathbf{I} - \mathbf{U}_k^H\mathbf{H}_k\mathbf{V}_k\right]^{-1}, \tag{4.10}$$

where $\beta = \sum_{i=1}^{K}\mathrm{Tr}\left(\frac{\sigma_i^2}{P_T}\mathbf{U}_i\mathbf{W}_i\mathbf{U}_i^H\right)$.

For the convenience of proof, replacing $\{\mathbf{V}_k\}$, $\{\mathbf{U}_k\}$, $\{\mathbf{W}_k\}$ in (4.8) by the precoding matrix $\mathbf{V}$, the receiving matrix $\mathbf{U}$, the weight matrix $\mathbf{W}$, respectively, we could obtain

$$\mathbf{V} = \left(\beta\mathbf{I} + \mathbf{H}^{\mathrm{H}}\mathbf{UWU}^{\mathrm{H}}\mathbf{H}\right)^{-1}\mathbf{H}^{\mathrm{H}}\mathbf{UW}. \tag{4.11}$$

As proven in [4], $\mathbf{W}$ is a Hermitian matrix, and thus there exists a matrix $\mathbf{Q}$ and a real diagonal matrix $\mathbf{\Lambda}$ such that $\mathbf{W} = \mathbf{Q}\mathbf{\Lambda}\mathbf{Q}^{\mathrm{H}}$. Substituting $\mathbf{W} = \mathbf{Q}\mathbf{\Lambda}\mathbf{Q}^{\mathrm{H}}$ into (4.11) results in

$$\begin{aligned}
\mathbf{V} &= \left(\beta\mathbf{I} + \mathbf{H}^{\mathrm{H}}\mathbf{UQ}\mathbf{\Lambda}\mathbf{Q}^{\mathrm{H}}\mathbf{U}^{\mathrm{H}}\mathbf{H}\right)^{-1}\mathbf{H}^{\mathrm{H}}\mathbf{UQ}\mathbf{\Lambda}\mathbf{Q}^{\mathrm{H}} \\
&= \mathbf{H}^{\mathrm{H}}\mathbf{UQ}\sqrt{\mathbf{\Lambda}}\left(\beta\mathbf{I} + \sqrt{\mathbf{\Lambda}}\mathbf{Q}^{\mathrm{H}}\mathbf{U}^{\mathrm{H}}\mathbf{HH}^{\mathrm{H}}\mathbf{UQ}\sqrt{\mathbf{\Lambda}}\right)^{-1}\sqrt{\mathbf{\Lambda}}\mathbf{Q}^{\mathrm{H}}.
\end{aligned} \tag{4.12}$$

Here, we define "$\mathbf{A} \rightarrow \mathbf{B}$" to denote that $\mathbf{B}$ is determined by $\mathbf{A}$. Pre-multiplying (4.12) by the channel matrix $\mathbf{H}$, we have

$$\mathbf{HH}^{\mathrm{H}} \rightarrow \mathbf{HV} \Leftrightarrow \mathbf{HH}^{\mathrm{H}} \rightarrow \left\{\mathbf{H}_i\mathbf{V}_j\right\}, \forall i, j. \tag{4.13}$$

Substituting (4.12) into $\mathbf{V}^{\mathrm{H}}\mathbf{V}$ results in

$$\mathbf{HH}^{\mathrm{H}} \rightarrow \mathbf{V}^{\mathrm{H}}\mathbf{V} \Rightarrow \mathbf{HH}^{\mathrm{H}} \rightarrow \left\{\mathrm{Tr}(\mathbf{V}_i\mathbf{V}_i^{\mathrm{H}})\right\}, \forall i. \tag{4.14}$$

Furthermore, based on (4.9) and (4.10), it is evident that

$$\left\{\mathbf{H}_i\mathbf{V}_j\right\}, \left\{\mathrm{Tr}(\mathbf{V}_i\mathbf{V}_i^{\mathrm{H}})\right\}, \forall i, j \rightarrow \left\{\mathbf{U}_k\right\}, \forall k, \tag{4.15}$$

$$\left\{\mathbf{H}_i\mathbf{V}_j\right\}, \left\{\mathrm{Tr}(\mathbf{V}_i\mathbf{V}_i^{\mathrm{H}})\right\}, \left\{\mathbf{U}_k\right\}, \forall i, j, k \rightarrow \left\{\mathbf{W}_k\right\}, \forall k. \tag{4.16}$$

Combining (4.13), (4.14), and (4.15), $\{\mathbf{U}_k\}$ is determined by $\mathbf{HH}^{\mathrm{H}}$, and considering (4.16), $\{\mathbf{W}_k\}$ is also determined by $\mathbf{HH}^{\mathrm{H}}$.

In conclusion, the receiving matrix $\mathbf{U}$ and the weight matrix $\mathbf{W}$ are solely determined by the channel correlation matrix $\mathbf{HH}^{\mathrm{H}}$.

4.3.2   Feature Extraction

Driven by the principle of high performance yet low complexity, the feature extraction module is designed based on the MLP-Mixer architecture. As depicted in Fig. 4.3a, the feature extraction module consists of one batch normalization layer, six Mixer layers, and a fully connected layer. Due to the influence of large-scale fading, the magnitude of the original channel matrix is very small, which can result in difficult training progress of the network. Therefore, we have incorporated a batch

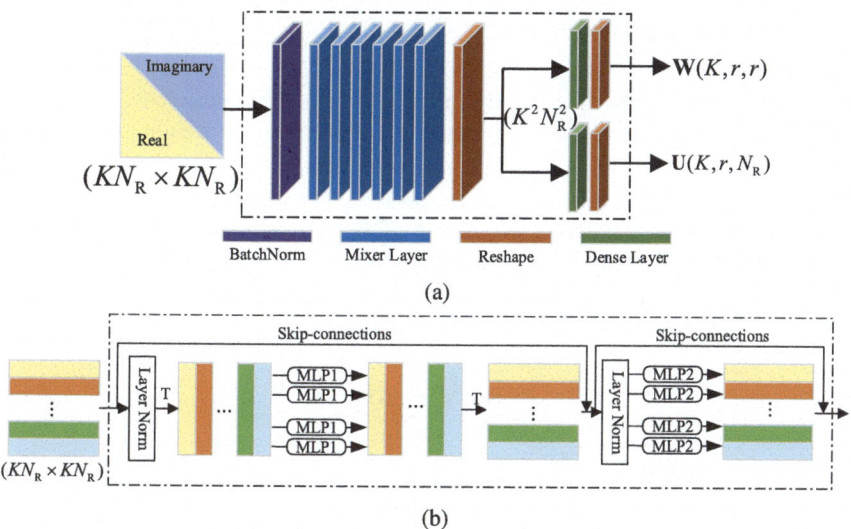

Fig. 4.3 Feature extraction module. (**a**) Overall structure. (**b**) Mixer layer structure

normalization layer at the beginning of the network, scaling the mean and variance of the data to a conventional range, which is more effective than manually designed scaling factors. Next, six layers of Mixer are utilized for feature extraction, which constitutes the core design of the network and will be discussed below in detail. Then, the output of the Mixer layer is flattened. Finally, after two fully connected layers and reshaping, $\mathbf{U}$ and $\mathbf{W}$ are obtained.

As illustrated in Fig. 4.3b, the data passes through layer normalization, token-mixing MLPs (MLP1), layer normalization, and channel-mixing MLPs (MLP2) sequentially in the Mixer layers. Layer normalization ensures a stable feature distribution for each sample. It effectively addresses training instability caused by differences in numerical magnitudes among different samples. MLP1 and MLP2 are responsible for mixing rows and columns, respectively. Each MLP is equipped with $4KN_R$ hidden neurons. The combination of MLP1 and MLP2 in a Mixer layer allows neurons located in the same row or column to jointly participate in feature computation, allowing for the full extraction of interrow correlations within channel matrix $\mathbf{H}$, namely, the correlations between channels of different antennas at the UE side. Furthermore, skip connections are incorporated in the Mixer layers to prevent degradation of network performance with increasing depth. Compared to blindly applying fully connected layers to all neurons, Mixer layer significantly reduces unnecessary parameters and computational overhead.

4.3.3 Precoding Recovery

As illustrated in Fig. 4.4, the recovery module consists of the pre-recovery submodule and the scaling submodule. The receiving matrix $\mathbf{U}$ and the weight matrix $\mathbf{W}$ are first transformed into the preliminary precoding matrices $\{\tilde{\mathbf{V}}_k\}$ and then scaled to the final precoding matrix $\mathbf{V}$.

Due to the large magnitude fluctuation of $\mathbf{H}_k$, directly using (4.8) as the pre-recovery module will cause large magnitude fluctuation of $\mathbf{W}_k$, which brings great difficulties to the network learning. Therefore, in pre-recovery submodule, the channel matrix of the kth UE is normalized to

$$\tilde{\mathbf{H}}_k \triangleq \frac{1}{\sqrt{q_k}}\mathbf{H}_k, \tag{4.17}$$

where $q_k \triangleq \mathrm{tr}(\mathbf{H}_k\mathbf{H}_k^{\mathrm{H}})$, so that the Frobenius norm of the normalized channel matrix of the kth UE $\tilde{\mathbf{H}}_k$ is 1. Accordingly, the noise power of the kth UE is scaled to

$$\tilde{\sigma}_k \triangleq \frac{1}{\sqrt{q_k}}\sigma_k. \tag{4.18}$$

Substituting (4.17) and (4.18) into (4.8), (4.9), and (4.10), it can be observed that the aforementioned approach is equivalent to scaling the receiving matrix of the kth UE $\mathbf{U}_k$, i.e., $\tilde{\mathbf{U}}_k = \sqrt{q_k}\mathbf{U}_k$, which could alleviate the scale fluctuation of elements of $\mathbf{U}_k$ caused by excessively large or small $\mathbf{H}_k$.

Therefore, replacing σ_k and $\mathbf{H}_k$ in (4.8) by $\tilde{\sigma}_k$ and $\tilde{\mathbf{H}}_k$, the preliminary precoding matrix of the kth UE $\tilde{\mathbf{V}}_k$ is given by

$$\tilde{\mathbf{V}}_k = \left[\sum_{i=1}^{K}\mathrm{Tr}\left(\frac{\tilde{\sigma}_i^2}{P_{\mathrm{T}}}\mathbf{U}_i\mathbf{W}_i\mathbf{U}_i^{\mathrm{H}}\right)\mathbf{I} + \tilde{\mathbf{H}}^{\mathrm{H}}\mathbf{UWU}^{\mathrm{H}}\tilde{\mathbf{H}}\right]^{-1}\tilde{\mathbf{H}}_k^{\mathrm{H}}\mathbf{U}_k\mathbf{W}_k, \forall k, \tag{4.19}$$

where $\tilde{\mathbf{H}} \triangleq [\tilde{\mathbf{H}}_1, \tilde{\mathbf{H}}_2, \cdots, \tilde{\mathbf{H}}_K]$. The precoding matrices $\tilde{\mathbf{V}}_k$ usually do not meet the power constraint, so the final precoding matrix of the kth UE is normalized to

$$\mathbf{V}_k = \sqrt{\frac{P_{\mathrm{T}}}{\sum_{i=1}^{K}\mathrm{tr}(\tilde{\mathbf{V}}_i\tilde{\mathbf{V}}_i^{\mathrm{H}})}}\tilde{\mathbf{V}}_k. \tag{4.20}$$

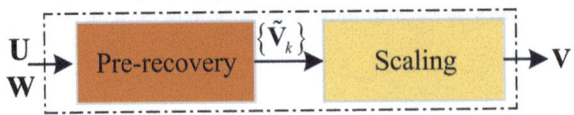

Fig. 4.4 Recovery module

4.3.4 Loss Function Design

Considering the high computational cost of WMMSE algorithm to obtain the near-optimal precoding, we adopt unsupervised training and design a loss function to avoid the requirement of label.

To be consistent with the recovery module, substituting (4.17) and (4.18) into (4.2), the sum rate can be rewritten as

$$R = \log\det\left[\mathbf{I} + \tilde{\mathbf{H}}_k \mathbf{V}_k \mathbf{V}_k^{\mathrm{H}} \tilde{\mathbf{H}}_k^{\mathrm{H}} \times \left(\sum_{i \neq k} \tilde{\mathbf{H}}_k \mathbf{V}_i \mathbf{V}_i^{\mathrm{H}} \tilde{\mathbf{H}}_k^{\mathrm{H}} + \tilde{\sigma}_k^2 \mathbf{I} \right)^{-1} \right]. \qquad (4.21)$$

To maximize the sum rate, the opposite number of sum rate is utilized as the loss function. Substituting (4.20) into (4.21), the loss function is given by

$$L = -\log\det\left[\mathbf{I} + \tilde{\mathbf{H}}_k \tilde{\mathbf{V}}_k \tilde{\mathbf{V}}_k^{\mathrm{H}} \tilde{\mathbf{H}}_k^{\mathrm{H}} \times \left(\sum_{i \neq k} \tilde{\mathbf{H}}_k \tilde{\mathbf{V}}_i \tilde{\mathbf{V}}_i^{\mathrm{H}} \tilde{\mathbf{H}}_k^{\mathrm{H}} + \sum_{j=1}^{K} \mathrm{tr}\left(\tilde{\mathbf{V}}_j \tilde{\mathbf{V}}_j^{\mathrm{H}} \right) \frac{\tilde{\sigma}_k^2}{P_{\mathrm{T}}} \mathbf{I} \right)^{-1} \right]. \qquad (4.22)$$

4.3.5 Generalization

In practical systems, the number of UEs served by the gNB could be changed, and the number of UE antennas may be different. However, the input and output sizes of the neural network are fixed. This raises a challenge for the application of the proposed precoding scheme.

To achieve compatibility with varying numbers of UEs and UE antennas, we can pretrain a neural network with a sufficient number of UEs and antennas, coupled with the zero-padding method to adapt to other scenarios. In cases where the UEs are fewer, zero-UEs (UEs with a channel matrix of $\mathbf{0}$) can be introduced to satisfy the input, and after computing $\mathbf{U}$ and $\mathbf{W}$, only the $\{\mathbf{U}_k\}$ and $\{\mathbf{W}_k\}$ of existing UEs are retained to reduce computational overhead in the precoding recovery module. Similarly, when the number of UE antennas is less, missing rows in the channel matrix can be filled with zero matrices as inputs, and the invalid portion of the network output should be discarded.

By zero padding, a single set of network parameters could accommodate various numbers of UEs and UE antennas, effectively reducing training costs and parameter storage space.

4.4   Performance and Discussions

In this section, the system parameters and dataset setup are first provided. Then, we compared the performance and computational complexity of the proposed network with traditional algorithms.

4.4.1   Simulation Setup

4.4.1.1   System Setup

As shown in Table 4.1, we consider two cases with different numbers of antennas and UEs. The noise power is set to -96.45 dBm, calculated based on a noise density of -174 dBm/Hz with an additional 5 dB of hardware noise.

The WINNER II model defined in 3GPP is adopted to generate the channel data [20]. 66,000 samples are considered in each case, with 60,000, 3000, and 3000 samples for training, validation, and testing, respectively.

The experiment parameters are listed in Table 4.2. In the training phase, the Graphics Processing Unit (GPU) is used to accelerate training, and in the test phase, only the Central Processing Unit (CPU) is employed to fairly compare the complexity. We adopt a learning rate decay strategy, with an initial learning rate of 10^{-4}. If the loss on the training set does not decrease within 20 consecutive epochs, the learning rate is halved.

Table 4.1 System parameters

Parameters	Case 1	Case 2
Subcarrier spacing	30 kHz	30 kHz
Number of resource blocks	50	50
N_T	16	64
N_R	2	2
K	8	24
r	1	1
P_T	53 dBm	53 dBm
σ_k^2	-96.45 dBm	-96.45 dBm

Table 4.2 Experiment parameters

Parameters	Value
Optimizer	Adam
Maximum number of epochs	750
Batch size	500
CPU	Intel Core i5-8265U
GPU	NVIDIA A16 15G

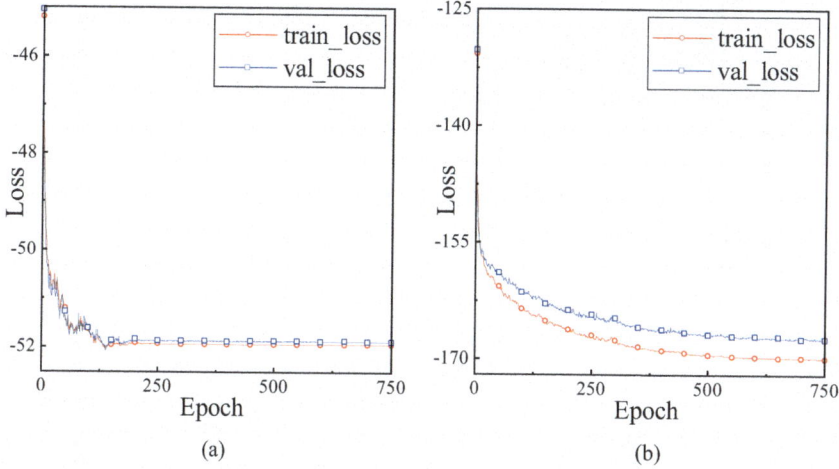

Fig. 4.5 Loss curves versus the number of training epochs. (**a**) Case 1. (**b**) Case 2

4.4.1.2 Benchmark Schemes

We compared the proposed MMPNet with some traditional baseline schemes: (a) high-performance baseline, i.e., WMMSE algorithm [4, 14]; (b) low-complexity baselines, i.e., BD algorithm and MET algorithm [8, 9].

4.4.2 Convergence Analysis

To verify the convergence of our proposed MMPNet model, the loss curves of case 1 and case 2 versus the number of training epochs are depicted in Fig. 4.5a, b, respectively. In both cases, the loss curves exhibit rapid descent within the first few epochs, followed by fluctuating declines until eventually stabilizing, which indicates model convergence.

4.4.3 Performance Evaluation

4.4.3.1 Generalization Performance Versus UE Number

The proposed model is trained with the maximum UE number and applied to other scenarios with the zero-padding method described in Sect. 4.3. As shown in Fig. 4.6, for both cases, MMPNet could achieve the performance close to WMMSE algorithm and outperform that of BD and MET algorithm, especially for the case with fewer UEs. As the number of UEs increases, the performance gap between

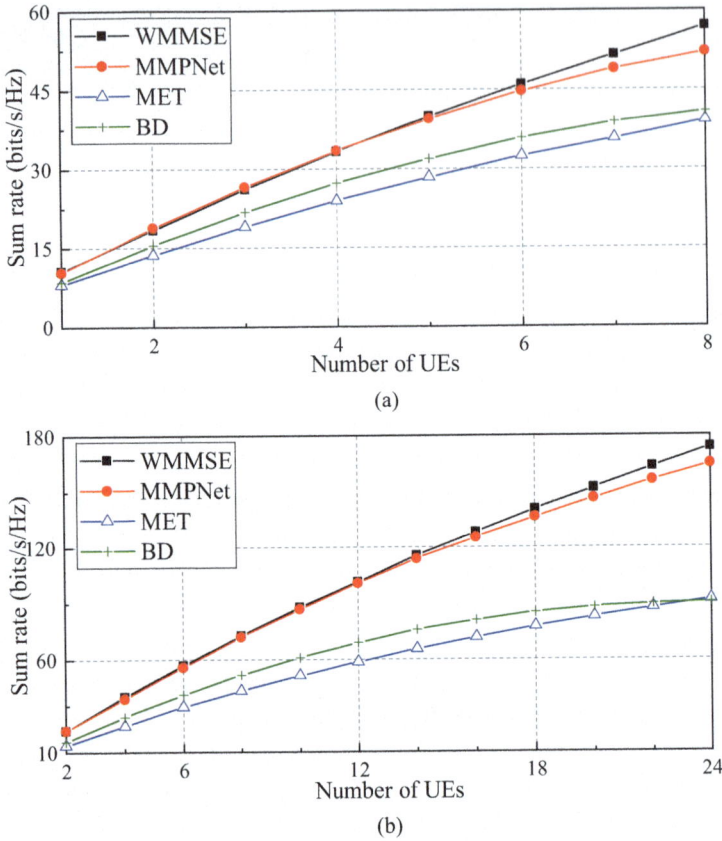

Fig. 4.6 Sum rate comparison. (**a**) Case 1. (**b**) Case 2

the WMMSE algorithm and MMPNet becomes larger. However, the gap falls in the acceptable ranges.

4.4.3.2 Generalization Performance Versus Transmit Power

To test the compatibility of MMPNet with different total maximum transmit power P_T, we directly applied the trained model to other scenarios with varying P_T. As illustrated in Fig. 4.7, when the transmit power is lower than the training power, the sum rate gap between MMPNet and WMMSE algorithm increases. Conversely, when the transmit power exceeds the training power, the gap decreases. It is noteworthy that the performance of the proposed MMPNet consistently outperforms that of BD and MET algorithm.

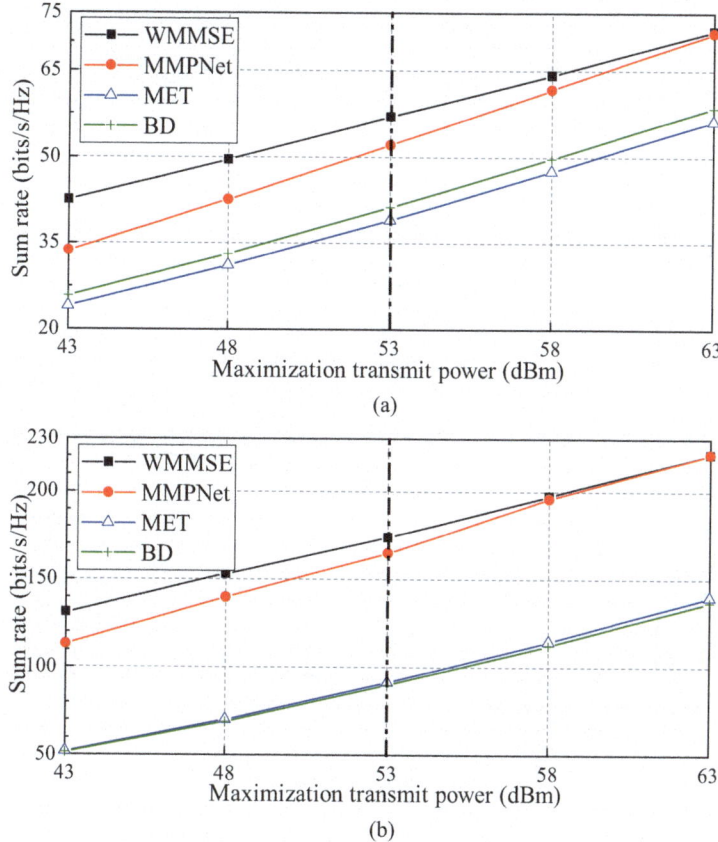

Fig. 4.7 Generalization performance versus the maximization transmit power. (**a**) Case 1. (**b**) Case 2

4.4.4   Complexity Evaluation

In this subsection, all test results are performed on the same CPU device, and the batch size is set to 1 for fairness. As shown in Fig. 4.8, WMMSE algorithm exhibits significantly longer runtime due to its extensive matrix inversion operations and iteration. The execution time of the proposed MMPNet has the same order of BD and MET algorithm and is significantly lower than that of WMMSE algorithm. The results also indicate that the runtime of MMPNet is nearly constant, as we use a single trained model to accommodate various numbers of UEs.

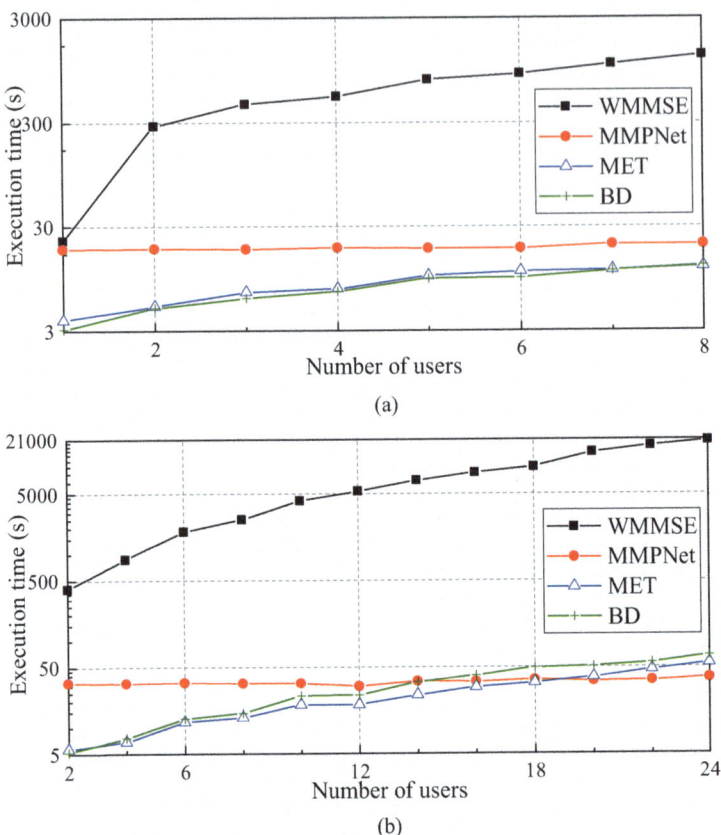

Fig. 4.8 Execution time comparison. (**a**) Case 1. (**b**) Case 2

4.5   Summary

This chapter proposed a low-complexity precoding model with unsupervised train-ing. The feasibility of using the compressed channel correlation matrix as input of the DL network was first proved. Subsequently, a low-complexity feature extraction module is designed to fit for the input structure, and the recovery module of the WMMSE algorithm is also improved. Finally, the scalability of the proposed model versus the varying numbers of UEs is discussed. Experimental results indicate that our proposed network model could achieve the close performance to WMMSE algorithm with significantly reduced computational complexity.

References

1. Ghosh, A., Maeder, A., Baker, M., Chandramouli, D.: 5G evolution: a view on 5G cellular technology beyond 3GPP release 15. IEEE Access **7**, 127639–127651 (2019)
2. de Figueiredo, F.A.P.: An overview of massive MIMO for 5G and 6G. IEEE Latin Am. Trans. **20**(6), 931–940 (2022)
3. Chen, W., et al.: 5G-advanced toward 6G: past, present, and future. IEEE J. Sel. Areas Commun. **41**(6), 1592–1619 (2023)
4. Shi, Q., Razaviyayn, M., Luo, Z.Q., He, C.: An iteratively weighted MMSE approach to distributed sum-utility maximization for a MIMO interfering broadcast channel. IEEE Trans. Signal Process. **59**(9), 3060–3063 (2011)
5. Chen, C.W., Tsai, W.C., Wong, S.S., Teng, C.F., Wu, A.Y.: WMMSE-based alternating optimization for low-complexity multi-IRS MIMO communication. IEEE Trans. Veh. Technol. **71**(10), 11234–11239 (2022)
6. Wang, K., Liu, A.: Robust WMMSE-based precoder with practice-oriented design for massive MU-MIMO . IEEE Wirel. Commun. Lett. **13**(7), 1858–1862 (2024)
7. Yoo, T., Goldsmith, A.: On the optimality of multiantenna broadcast scheduling using zero-forcing precoding. IEEE J. Sel. Areas Commun. **24**(3), 528–541 (2006)
8. Spencer, Q.H., Swindlehurst, A.L., Haardt, M.: Zero-forcing methods for downlink spatial multiplexing in multiuser MIMO channels. IEEE Trans. Signal Process. **52**(2), 461–471 (2004)
9. Boccardi, F., Huang, H.: A near-optimum technique using linear precoding for the MIMO broadcast channel. In: IEEE International Conference on Acoustics, Speech and Signal Processing, pp. 17–20 (2007)
10. Sandell, M., Vetter, H., Tosato, F.: Joint linear and nonlinear precoding in MIMO Systems. IEEE Commun. Lett. **15**(12), 1265–1267 (2011)
11. Bai, X., Peng, Q.: An online deep learning based channel estimation method for mmWave massive MIMO systems. In: 2023 IEEE 97th Vehicular Technology Conference (VTC2023-Spring), pp. 1–5 (2023)
12. Shen, H., Zhao, L., Wang, F., Cao, Y.: MixerNet: deep learning for eigenvector-based CSI feedback. In: 2022 14th International Conference on Wireless Communications and Signal Processing (WCSP), pp. 1167–1171 (2022)
13. Olatinwo, S.O., Joubert, T.-H.: Deep learning for resource management in internet of things networks: a bibliometric analysis and comprehensive review. IEEE Access **10**, 94691–94717 (2022)
14. Hu, Q., Cai, Y., Shi, Q., Xu, K., Yu, G., Ding, Z.: Iterative algorithm induced deep-unfolding neural networks: precoding design for multiuser MIMO systems. IEEE Trans. Wirel. Commun. **20**(2), 1394–1410 (2021)
15. Lu, S., Zhao, S., Shi, Q.: Learning-based massive beamforming. In: GLOBECOM 2020—2020 IEEE Global Communications Conference, pp. 1–6 (2020)
16. Zhang, S., Xu, J., Xu, W., Wang, N., Ng, D.W.K., You, X.: Data augmentation empowered neural precoding for multiuser MIMO with MMSE model. IEEE Commun. Lett. **26**(5), 1037–1041 (2022)
17. Xia, W., Zheng, G., Zhu, Y., Zhang, J., Wang, J., Petropulu, A.P.: A deep learning framework for optimization of MISO downlink beamforming. IEEE Trans. Commun. **68**(3), 1866–1880 (2020)
18. Liu, F., Zhang, L., Du, R., Li, D., Li, T.: Two-stage hybrid precoding for minimizing residuals using convolutional neural network. IEEE Commun. Lett. **25**(12), 3903–3907 (2021)
19. Tolstikhin, I., Houlsby, N., Kolesnikov, A., et al.: MLP-Mixer: An all-MLP architecture for vision (2021). https://arxiv.org/abs/2105.01601
20. 3GPP: Study on channel model for frequencies from 0.5 to 100 GHz. 3GPP (2022)

Chapter 5
Intelligent Beam Management Technology

Abstract The excessive overhead of millimeter-wave beam measurement induces a significant challenge for massive MIMO systems. To reduce the measurement overhead and improve the accuracy of beam selection, DL is employed to predict the signal quality of all beam pairs by measuring a few selected beam pairs in this chapter. The proposed beam prediction network is mainly composed of input module, residual module, and MLP module, therefore denoted as ReMBP net. Moreover, to improve the generalization ability of the proposed ReMBP net for different communication environments, the input module with a flexible input size of signal quality data is designed.

Keywords AI · Millimeter-wave · MIMO · Beam management · Beam prediction

5.1 Background

In the 5G mobile communication systems, massive MIMO technology is employed to form high-gain analog beams, therefore compensate for the large path loss caused by millimeter wave, and ensure the system coverage [1–3]. However, the beam management should be adopted in order to obtain the optimal beam pair between the next gNB and UE. The process of beam management can be divided into beam sweeping, beam measurement, beam determination, and beam reporting [4, 5]. At the beam sweeping stage, the overhead and delay significantly increase with the increasing number of antennas in massive MIMO systems, such as 2048 beam pairs that are required to be measured when the gNB with 64 antennas communicates with a UE with 32 antennas. Therefore, how to select the best beam pair among these beam pairs formed by the gNB and UE while guaranteeing low measurement overhead becomes one hot research field.

Traditionally, the gNB and UE sweep all the beam pairs during the initial access, and then the UE determines the optimal beam pair based on its measured signal quality, such as Reference Signal Received Power (RSRP), Reference Signal Receiving Quality (RSRQ), or SNR [6, 7]. However, the measurement overhead of traditional method is too large to be applied for practical MIMO systems. Therefore, a hierarchical beam sweeping scheme is proposed to alleviate the excessive measurement overhead [8]. The hierarchical scheme mainly contains two steps: A wide beam sweeping is first carried out, and the beam index with the highest signal strength is selected by the UE and sent to the gNB; then the gNB sweeps multiple narrow beams based on the feedback information, and the UE reports the best beam pair; the system could repeat the two steps until the signal gain of the selected beam pair satisfies the system requirement. Although these traditional beam management schemes mentioned above could save some measurement overhead, they are still insufficient for mmWave MIMO systems.

On the other hand, with the stunning breakthroughs of calculating capacity and the expansion of data scale, AI technology has developed rapidly, and DL algorithms have made a lot of progress in driving decision-making, Computer Vision (CV), etc. [9]. The combination of AI and wireless communication systems brings new solutions to traditional communication problems. For example, DL algorithms can accurately model complex features and realize super-resolution beam prediction [10]. Some AI methods have been introduced into beam management [6, 11, 12]. A DL model composed of 13 convolutional layers is designed to judge the channel type (LOS or NLOS) and then select the suitable beam based on the channel type [11]. The Original DNN-based Beam Training (ODBT) and Enhanced DNN-based Beam Training (EDBT) are studied for beam prediction, and the measurement overhead can be reduced by 53% compared with the traditional beam sweeping scheme [12]. Then, a multimodal Machine Learning (ML)-based method could rapidly predict the beam direction in mmWave communication systems [13]. A model based primarily on meta-learning is achieved to facilitate more accurate beam prediction [14]. And a DL model utilizes radar perception data to guide millimeter-wave beam prediction, significantly reducing the beam measurement overhead [15]. However, the measurement overhead of the current studies still needs to be reduced for massive MIMO systems, the performance of beam management should be further improved, and an AI-based beam management scheme for flexible scenarios that balances system overhead and performance should be studied for practical systems. Meanwhile, most of these studies focused primarily on spatial beam prediction, with insufficient consideration given to the temporal correlation of beams. A comprehensive model should be capable of simultaneously extracting both the spatial and temporal characteristics of the beams.

Inspired by the above investigation, to improve the accuracy of beam selection and reduce the overhead of beam measurement, a new ReMBP net for beam selection is proposed. The ReMBP net is mainly composed of input module, residual module, and MLP module. Among all the beam pairs between the gNB and UE, the RSRP values of a few measured beam pairs are first selected as the input information of the proposed ReMBP net to reduce the system overhead. The input RSRP data is first preprocessed by the input module; then, the residual module is adopted to extract the high-dimensional feature vector; the RSRP results of all beam pairs are finally predicted through a designed MLP module. On the other hand, the flexible input size of RSRP data is studied in order to support various scenarios for practical systems.

5.2   System Model and Problem Formulation

5.2.1   System Model

In 5G massive MIMO systems, the downlink beam management is considered with the Three-Dimensional (3D) millimeter-wave channels, where the antennas are placed in Uniform Planar Arrays (UPAs). As shown in Fig. 5.1, the gNB is equipped with an array of $N_{\mathrm{Th}} \times N_{\mathrm{Tv}} (= N_{\mathrm{T}})$ antennas in the horizontal and vertical directions; similarly, the UE is equipped with an array of $N_{\mathrm{Rh}} \times N_{\mathrm{Rv}} (= N_{\mathrm{R}})$ antennas in the horizontal and vertical directions. The beams formed by the arrays at the gNB and UE together constitute the beam codebook $\tilde{Q}$ with the size of $N_{\mathrm{T}} \times N_{\mathrm{R}}$, which can be expressed by

$$\tilde{Q} = \{(\mathbf{f}_1, \mathbf{w}_1), (\mathbf{f}_1, \mathbf{w}_2), \cdots, (\mathbf{f}_{N_{\mathrm{T}}}, \mathbf{w}_{N_{\mathrm{R}}})\}, \tag{5.1}$$

where $\mathbf{f}_i$ ($i \in \{1, 2, \cdots, N_{\mathrm{T}}\}$) and $\mathbf{w}_j$ ($j \in \{1, 2, \cdots, N_{\mathrm{R}}\}$) represent the transmit beam and receive beam, respectively.

Assuming that the number of multipaths is P between the gNB and UE, as well as β_p, $\theta_{\mathrm{h},p}$, $\theta_{\mathrm{v},p}$, $\phi_{\mathrm{h},p}$, and $\phi_{\mathrm{v},p}$ denote the fading coefficient, Azimuth of Arrival (AoA), zenith AoA, Azimuth of Departure (AoD), and zenith AoD of the pth channel path [16], then the channel matrix for 3D massive MIMO can be expressed as

$$\mathbf{H} = \sqrt{\frac{N_{\mathrm{T}} N_{\mathrm{R}}}{P}} \sum_{p=1}^{P} \beta_p \mathbf{a}_{\mathrm{r}}^{\mathrm{T}} \left(\theta_{\mathrm{h},p}, \theta_{\mathrm{v},p} \right) \mathbf{a}_{\mathrm{t}} \left(\phi_{\mathrm{h},p}, \phi_{\mathrm{v},p} \right), \tag{5.2}$$

where the steering vector

$$\mathbf{a}_{\chi} \left(\vartheta_{\mathrm{h}}, \vartheta_{\mathrm{v}} \right) = \mathbf{a}_{\mathrm{h}} \left(\vartheta_{\mathrm{h}} \right) \otimes \mathbf{a}_{\mathrm{v}} \left(\vartheta_{\mathrm{v}} \right), \tag{5.3}$$

$$\mathbf{a}_{\mathrm{h}} \left(\vartheta_{\mathrm{h}} \right) = \frac{1}{\sqrt{N_{\mathrm{h}}}} \left[1, e^{j\pi \sin(\vartheta_{\mathrm{h}})}, \cdots, e^{j(N_{\mathrm{h}}-1)\pi \sin(\vartheta_{\mathrm{h}})} \right]^{\mathrm{T}}, \tag{5.4}$$

Fig. 5.1 Illustration of multiple beams at the gNB and UE in millimeter-wave massive MIMO systems

$$\mathbf{a}_v\left(\theta_v\right) = \frac{1}{\sqrt{N_v}}\left[1, e^{j\pi\sin(\theta_v)}, \cdots, e^{j(N_v-1)\pi\sin(\theta_v)}\right]^{\mathsf{T}}, \tag{5.5}$$

$$\chi \in \{r, t\}, \vartheta \in \left\{\theta_{h,p}, \phi_{h,p}\theta_{v,p}, \phi_{v,p}\right\}. \tag{5.6}$$

Suppose that x denotes the transmitted modulation symbol, and then the received signal after received beamforming can be expressed as

$$y_{ij} = \mathbf{w}_j^{\mathsf{H}}\mathbf{H}\mathbf{f}_i x_{ij} + \mathbf{w}_j^{\mathsf{H}}\mathbf{n}_{ij}, \tag{5.7}$$

where $\left(\mathbf{w}_i, \mathbf{f}_j\right) \in \tilde{Q}$ and $\mathbf{n}_{ij} \sim CN\left(0, \sigma^2\mathbf{I}\right)$ denotes the Additive White Gaussian Noise (AWGN) with zero mean and the variance of σ^2.

5.2.2 AI-Based Beam Management

In traditional beam management, the gNB and UE should sweep all the beams in different time slots and obtain the RSRP values, and then the beam pair with the best RSRP is adopted for the communication link. However, the time overhead ($\propto N_T \times N_R$) increases with the numbers of beams at the gNB and UE; therefore, it is not feasible for practical massive MIMO systems. To reduce the measurement overhead of beam pairs, the AI-based beam management is studied. In AI-based scheme, a few beam pairs are selected for RSRP measurement, and the RSRP values of other beam pairs could be predicted by AI model; then the best beam pair could be determined by the predicted RSRP values.

Supposing that M_T beams at the gNB and M_R beams at the UE are selected for RSRP measurement, then the selected beam codebook can be expressed as

$$Q = \{(\mathbf{f}_1, \mathbf{w}_1), (\mathbf{f}_1, \mathbf{w}_2), \cdots, (\mathbf{f}_{M_T}, \mathbf{w}_{M_R})\} \subset \tilde{Q}. \tag{5.8}$$

Based on the selected beam codebook Q, the dataset of RSRP value can be calculated as

$$C = \left\{r_{ij} = \|y_{ij}\|^2 | i \in \{1, 2, \cdots, M_T\}, j \in \{1, 2, \cdots, M_R\}\right\}. \tag{5.9}$$

By assuming that the proposed ReMBP net as $f_{\mathbf{W}}(\cdot)$, where $\mathbf{W}$ represents the training weight matrix, then we can obtain the RSRP values of all beam pairs, i.e.,

$$\tilde{C} = f_{\mathbf{W}}(C) \in R^{N_T \times N_R}. \tag{5.10}$$

Based on predicted RSRP values, we could obtain the optimal beam pair. However, to avoid the prediction error, the N beam pairs with best RSRP values in $\tilde{Q}$ are reported to the gNB. The AI model should guarantee that the predicted N

beam pairs contain the optimal beam pair. Then, the gNB could sweep the N beam pairs, as well as the UE can measure their RSRP values and report the beam pair with the best RSRP to the gNB. Therefore, the optimal beam pair can be obtained with extremely high probability.

5.2.3 Problem Formulation

Based on the procedure of the AI-based beam management, only $M_T \times M_R$ beam pairs for initial RSRP measurement and N beam pairs for secondary RSRP measurement need to be swept for the system. Therefore, the overhead that compared to traditional scheme can be calculated by the measure ratio, i.e., the number of measured beam pairs to the number of all beam pairs, which can be written as

$$\eta = \frac{M_T M_R + N}{N_T N_R}. \tag{5.11}$$

However, we should notice that the actual measure ratio may be smaller than η, because the predicted N beam pairs with best RSRP values may be the same as the initial measured beam pairs that have already been measured.

The number N of beam pairs for secondary RSRP measurement is dependent on the accuracy of AI model. On the other hand, the accuracy is dependent on the number of beam pairs for initial RSRP measurement. Therefore, the objective of this chapter is to minimize the average error between the predicted RSRP $\hat{r}$ and the measured RSRP r by optimizing the weight matrix $\mathbf{W}$ of the proposed AI model, i.e.,

$$\mathbf{W} = \arg \min_{\mathbf{W}} \left\{ \frac{1}{N_T \times N_R} \sum_{n=1}^{N_T \times N_R} |r_n - \hat{r}_n|^2 \right\}. \tag{5.12}$$

Meanwhile, we should try to minimize the number $M_T \times M_R$ of beam pairs for initial RSRP measurement or could adjust that number for different communication environments to guarantee the communication quality.

5.3 ReMBP Net for Spatial Beam Prediction

The specific structure of the proposed ReMBP net and the functions of the corresponding modules are first introduced in this section. Then, the beam prediction scheme is designed for a fixed number of measured beam pairs, denoted as ReMBP-a net. Moreover, the beam prediction with a flexible number of measured beam pairs is proposed to improve the generalization ability for different communication environments, denoted as ReMBP-b net. Finally, the loss function is designed for beam selection.

5.3.1 Structure of Proposed ReMBP Net

As shown in Fig. 5.2, the proposed ReMBP net includes input module, residual module, and MLP module, which will be described as follows in detail.

1. *Input module*: The main function of input module is to preprocess the input data, and the module is composed of convolutional layer, Batch Normalization (BN) layer, and LeakyReLU activation function. The convolutional layer is composed of a 2D convolution with kernel size of 1×1, which is used to increase the number of feature map channels without changing the size of input data. The BN layer is adopted to solve the problem that the data distribution of the middle layer changes during the training process, so as to prevent the gradient from disappearing or exploding, as well as accelerate the training speed and prevent overfitting. The LeakyReLU function is employed to introduce nonlinear factors and therefore improve the expression ability of the model. Different from the traditional Rectified Linear Unit (ReLU) function, the LeakyReLU function assigns a nonzero slope to the negative value, which could fix some parameters and eliminate gradient death [17]. Denote x as the input data, and then the LeakyReLU activation function can be expressed as

$$y = \max(0, x) + \varepsilon \times \min(0, x), \tag{5.13}$$

where ε is a very small constant and its default value is set as 0.05.

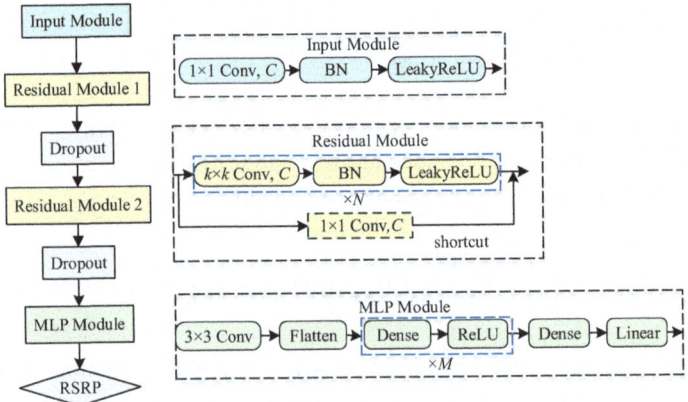

Fig. 5.2 Structure of the proposed ReMBP net

2. *Residual module*: In traditional neural networks, the gradient drop or gradient explosion could happen with the increasing number of network layers, which makes the model performance tend to be saturated or even deteriorate rapidly. Therefore, the proposed model introduces the residual block to alleviate the performance degradation caused by deepened network [18]. Suppose that the input of the residual module is x^l, and the output of the whole residual model can be expressed as $H(x^l)$, which contains two branches. As shown in Fig. 5.2, through a convolutional layer, the input data becomes $\sigma^{(l+1)}$, and then the output after BN layer and LeakyReLU function is $x^{(l+1)}$; by iteration of N times, the whole output of the first branch is denoted as $x^{(l+N)}$. On the other hand, the original input x^l is retained by the shortcut branch with the convolutional kernel size of 1×1, where the convolution operation is adopted to keep the output channels of both branches consistent.

By observing the structure of the residual model, we can see that the output feature of the later layer will be contributed by the previous layer. Therefore, if the network layer of the first branch is not effective, it can be skipped to ensure that the network performance will not deteriorate at least. In the other words, the second branch in the residual model could solve the problem of low learning efficiency and failure to effectively improve the accuracy.

3. *MLP module*: After the residual module 2, the feature map passes through the dropout layer in order to randomly discard some network parameters, which could prevent the model from overfitting. Then, a convolutional layer with a stride of 2 is adopted to compress the feature map size. Moreover, the flatten layer is employed to expand feature map into a one-dimensional feature vector, which is the input of the MLP module. The MLP module includes multiple combinations of dense layers and ReLU, and then the output of MLP is obtained through a dense layer and a linear activation function. Therefore, the input features of MLP module are adjusted by weights and deviations, as well as the predicted RSRP set $\tilde{C}$ of beam codebook $\tilde{Q}$ is finally obtained.

5.3.2 Beam Prediction and Selection Based on ReMBP Net

In order to reduce the measurement overhead of massive MIMO systems, we divide all beam pairs in beam codebook $\tilde{Q}$ into measurement beam pairs and non-measurement beam pairs. The RSRP values of M_T beams at the gNB and M_R beams at the UE are selected, and therefore the size of input data is $M_T \times M_R$.

For the fixed number of measured beam pairs, ReMBP-a net is designed based on ReMBP net. However, for practical systems, the number of measurement beam pairs needs to adapt different UE positions or communication environments and ensure the quality of the communication link. Therefore, the flexible number of measured beam pairs should be supported, which results in different input dimensions of AI model. Hence, we designed a ReMBP-b net for the flexible number of measured beam pairs.

1. Beam prediction for a fixed number of measured beam pairs
   ReMBP-a net is first designed for the fixed number $M_T \times M_R$ of measured beam pairs in single communication environment. To facilitate convolution operation, one data dimension is first added to the input data size of (M_T, M_R), and it becomes $(M_T, M_R, 1)$ as input. Then, the input module increases the number of channels so that it generates the feature map with a size of $(M_T, M_R, 64)$. The kernel size $k = 3$ is adopted in both residual module 1 and residual module 2. The number C of channels is set as 64 in residual module 1, therefore the feature map size is unchanged, and no convolutional layer is needed in the shortcut connection. In residual module 2, the number C of channels is increased to 128, the size of output feature map becomes $(M_T, M_R, 128)$, and therefore the number C of channels in the shortcut is set as 128. Moreover, after the convolutional layer with a stride of 2, the feature map becomes $(M_T/2, M_R/2128)$. Finally, the feature map is flattened to be a one-dimensional feature vector with size of $32 \times M_T \times M_R$, which is the input of the MLP module.

2. Beam prediction for a flexible number of measured beam pairs
   In order to satisfy the flexible number of measured beam pairs for different environments, ReMBP-b net is proposed by redesigning the input dataset. The input size of model is set as the size of beam codebook $\tilde{Q}$, i.e., $N_T \times N_R$. However, only the measured beam pairs with a flexible size of $M_T \times M_R$ have their measured RSRP values, and the RSRP values of non-measured beam pairs are set as a constant Ω, such as 0. Therefore, the dataset could be composed of different sizes of $M_T \times M_R$ as input data.

   For the input module, the number of channels increases from 1 to 64 so that the feature map becomes $(N_T, N_R, 64)$. Then, to expand the range of receptive fields, the kernel size of $k = 5$ is adopted in residual module 1. Because the RSRP values of non-measured beam pairs are set as a constant Ω in dataset, a lot of redundant information exists, and therefore an average pooling layer with a size of 2 is adopted to reduce the size of feature map to $(N_T/2, N_R/2, 64)$. For the residual module 2, the size of convolutional kernel is set as $k = 3$, and the number of channels is increased to $C = 128$, and therefore, the feature map becomes $(N_T/2, N_R/2128)$; similarly, an average pooling layer with a size of 2 is adopted to reduce the size of feature map to $(N_T/4, N_R/4128)$. Finally, the feature map is flattened to a one-dimensional feature vector with a size of $8 \times N_T \times N_R$ and then is the input to the MLP module.

5.3.3   Loss Function Design

The optimization objective of the ReMBP net is to minimize the average error between the predicted RSRP value and the measured RSRP value. Therefore, the

Mean Square Error (MSE) is taken as the loss function, which can be expressed as

$$L_{\text{MSE}} = \sum_{n=1}^{N_{\text{T}} \times N_{\text{R}}} (r_n - \hat{r}_n)^2, \tag{5.14}$$

where r_n and $\hat{r}_n$ denote the measured RSRP value and the predicted RSRP value, respectively.

However, based on the description of system model in Sect. 5.2.1, the prediction accuracy of the optimal N beam pairs is more important for practical systems. Therefore, we add an extra loss $L_{\text{Top}-N}$ to ensure the prediction accuracy for the optimal N beam pairs, and the extra loss can be expressed as

$$L_{\text{Top}-N} = \frac{1}{N} \sum_{n=1}^{N} (r_n - \hat{r}_n)^2. \tag{5.15}$$

Hence, the two loss functions are added as a joint loss function to optimize the net parameters during the training, i.e., the total loss can be expressed as

$$L_{\text{J}-\text{MSE}} = L_{\text{MSE}} + \gamma L_{\text{Top}-N}, \tag{5.16}$$

where N and γ denote the number of optimal beam pairs that needs to be reported to the gNB and the regularization coefficient that adjusts the proportion of $L_{\text{Top}-N}$.

5.4 Performance Evaluation

This section first introduces the dataset and parameter setup for experiments, and then the experiment results and analysis are given.

5.4.1 Dataset and Simulation Parameters Setup

The dataset is generated by system-level simulation platform. The gNB with 128 antennas has 64 beams, which are composed of 16 beams in the horizontal direction and 4 beams in the vertical direction. The UE is configured with three arrays, and the optimal array is selected in advance for practical applications; each array has eight antennas and forms four beams. We generated 39,600 samples in dataset, 95% of which is selected as the training set and 5% is selected as the test set.

The main parameter setup for model training is given in Table 5.1. During the training, loss value is employed as indicator, and the training is terminated when the loss does not decrease during 40 consecutive epochs, while the model parameters

Table 5.1 Setup of experiment parameters

Parameter	Value
Initial learning rate	0.001
Maximum number of epoches	500
Optimizer	Adam
Batch size	512
Early stop tolerance	40
Graphics card	Tesla V100 32G

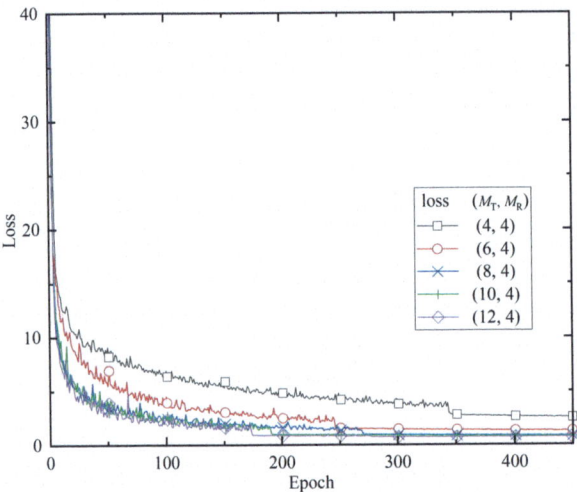

Fig. 5.3 Loss function for different numbers of measured beam pairs

with the minimum loss value are saved by orderly comparing the results of different epochs before stopping training. The initial learning rate is set as $l_r = 0.001$, and it will become $l_r = 0.1 \times l_r$ during the training process, if the loss value dose not improve within 30 consecutive epochs.

5.4.2 Results and Analysis

1. Results of ReMBP-a Net

   The convergence curves of our proposed ReMBP-a net are shown in Fig. 5.3. It can be seen that the loss value rapidly decreases in the initial stage and then becomes slow. However, some stair drops can be observed because the learning rate adjustment is adopted in the middle stage. Finally, all the loss curves converge with the increasing number of epochs. Moreover, the loss curve with a large number of measured beam pairs converges to a lower value.

   When $N_T = 64$, $N_R = 4$, and $M_R = 4$, the simulation results of proposed ReMBP-a net and VGG16 [19] for different numbers M_T of transmit

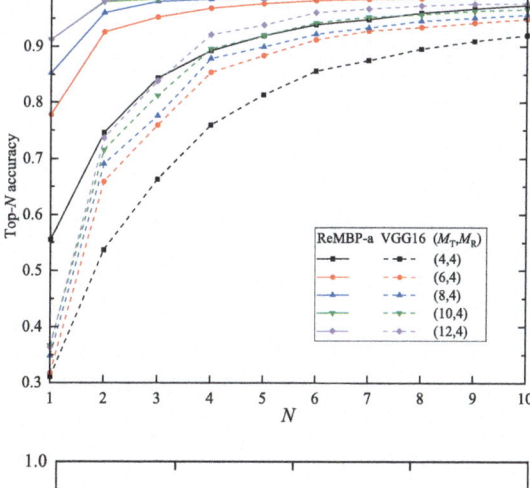

Fig. 5.4 Top-N accuracy of ReMBP-a with different M_T

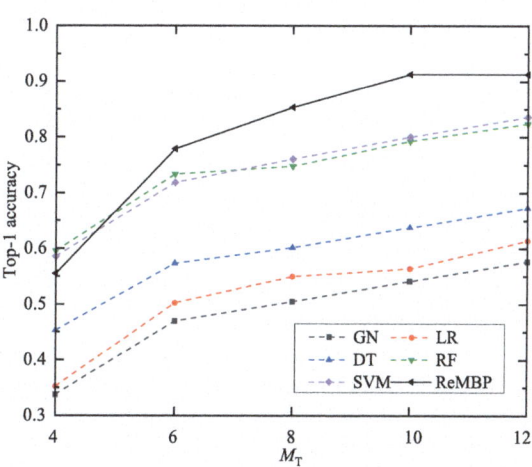

Fig. 5.5 Performance comparison of ReMBP-a net with traditional ML algorithms

beams are shown in Fig. 5.4. Given a fixed N, the beam selection accuracy becomes better with the increasing number M_T of transmit beams for both the ReMBP-a net and VGG16, because the AI network could learn more space-time information to improve the accuracy of predicted RSRP values. With the increasing number N of the secondary measured beam pairs, the top-N accuracy gradually increases, which is consistent with common sense. Moreover, we can see that the performance of the proposed ReMBP-a net is obviously better than that of the VGG16 net, which shows the advantage of our proposed model.

The Top-1 accuracy of proposed ReMBP-a net versus the number $M_T \times 4$ of measured beam pairs is illustrated in Fig. 5.5 compared with five traditional ML algorithms, i.e., Support Vector Machine (SVM), Random Forest (RF), Decision Tree (DT), Logistic Regression (LR), and Gaussian Naive Bayes (GNB). When the number of measured beam pairs is $M_T = 4$, the Top-1 accuracy of ReMBP-a net is about 4% lower than that of SVM and RF; however, when M_T becomes 6,

Table 5.2 Measurement
overhead η with accuracies of
95% and 98%

(M_T, M_R)	η @95%	η @98%
(12,4)	19.53%	19.53%
(10,4)	16.41%	16.80%
(8,4)	13.28%	13.67%
(6,4)	10.55%	11.72%
(4,4)	9.38%	10.55%

Fig. 5.6 Top-N accuracy
under the flexible number of
measured beam pairs

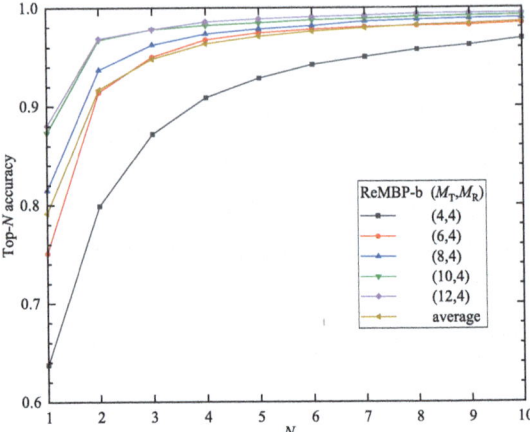

8, 10, or 12, the performance of ReMBP-a net is significantly improved than the five traditional ML methods.

On the other hand, to show the measurement overhead of our proposed model, the measure ratios with accuracies of 95% and 98% are listed in Table 5.2 for different numbers of measurement beam pairs. It can be seen that the measure ratio with $(M_T, M_R) = (4, 4)$ beam pairs is lowest under the two required accuracy ratios. Therefore, under the required accuracy ratios of 95% or 98%, the scheme with $(M_T, M_R) = (4, 4)$ is recommended for practical systems.

2. Results of ReMBP-b net

The experiment results for the flexible number (N_T, N_R) of measured beam pairs are given in Fig. 5.6. In the dataset, different numbers M_T of transmit beams are selected for RSRP measurement, as well as 4, 6, 8, 10, and 12 are adopted in our experiments.

With the increase of N, the Top-N accuracy of different numbers of measurement beam pairs is improved. Similarly, the beam selection accuracy becomes better with the increasing number M_T of transmit beams for the ReMBP-b net. Compared with the results under the fixed number of measured beam pairs in Fig. 5.4, the Top-1 accuracy under a flexible number of measured beam pairs is improved by 8% when $M_T = 4$. While for other values of M_T, the Top-1 accuracy under the flexible number of measured beam pairs becomes worse than that under the fixed number of measured beam pairs. This means that the ReMBP-b net faces greater difficulties in extracting the space-time feature information from the training dataset compared with the network trained separately for each scenario.

5.5 Summary

Beam management is a fundamental issue in mmWave communication systems. A key challenge in beam management is to reduce the spatial overhead associated with beam selection. Hence, the ReMBP net is first proposed to extract the features of a few RSRP values of measured beam pairs through residual module and then predict the RSRP values of all beam pairs. Therefore, the best beam pair with the best RSRP can be obtained for communication link. Moreover, the proposed ReMBP net is improved for the flexible size of input RSRP values of beam pairs to satisfy the requirements of different communication environments. Guaranteeing the accuracy of beam pair selection and prediction, experiment results indicate that the beam measurement overhead can be significantly reduced for the ReMBP net compared with VGG16 and some traditional ML algorithms.

References

1. Ramireddy, V., Grossmann, M., Landmann, M., Del Galdo, G.D.: Sub-band versus space-delay precoding for wideband mmWave channels. IEEE Wirel. Commun. Lett. **8**(1), 193–196 (2019)
2. Rappaport, T.T.S., Sun, S., Mayzus, R., Zhao, H., Azar, Y., Wang, K., Wong, G.N., Schulz, J.K., Samimi, M., Gutierrez, F.: Millimeter wave mobile communications for 5G cellular: It will work! IEEE Access **1**, 335–349 (2013)
3. Roh, W., et al.: Millimeter-wave beamforming as an enabling technology for 5G cellular communications: theoretical feasibility and prototype result. IEEE Commun. Mag. **52**(2), 106–113 (2014)
4. Giordani, M., Polese, M., Roy, A., Castor, D., Zorzi, M.: A tutorial on beam management for 3GPP NR at mmWave frequencies. IEEE Commun. Surv. Tutorials **21**(1), 173–196 (2019)
5. Giordani, M., Mezzavilla, M., Barati, C.N., Rangan, S., Zorzi, M.: Comparative analysis of initial access techniques in 5G mmWave cellular networks. In: Annual Conference on Information Systems and Sciences (CISS), pp. 268–273 (2016)
6. Giordani, M., Mezzavilla, M., Zorzi, M.: Initial access in 5G mmWave cellular networks. IEEE Commun. Mag. **54**(11), 40–47 (2016)
7. Jeong, C., Park, J., Yu, H.: Random access in millimeter-wave beamforming cellular networks: issues and approaches. IEEE Commun. Mag. **53**(1), 180–185 (2015)
8. Desai, V., Krzymien, L., Sartori, P., Xiao, W., Soong, A., Alkhateeb, A.: Initial beamforming for mmWave communications. In: 48th Asilomar Conference on Signals, Systems and Computers, pp. 1926–1930 (2014)
9. Wang, Z., Gao, P., He, Z., Zhao, L.: A CGAN-based model for Human-like driving decision making. In: IEEE Wireless Communications and Networking Conference (WCNC), pp. 1926–1930 (2021)
10. Ma, K., Wang, Z., Tian, W., Chen, S., Hanzo, L.: Deep Learning for mmWave beam-management: state-of-the-art, opportunities and challenges. IEEE Wirel. Commun. **30**(4), 108–114 (2023)

11. Klautau, A., González-Prelcic, N., Heath, R.W.: LIDAR data for deep learning-based mmWave beam-selection. IEEE Wirel. Commun. Lett. **8**(3), 909–912 (2019)
12. Qi, C., Wang, Y., Li, G.Y.: Deep learning for beam training in millimeter wave massive MIMO systems. IEEE Trans. Wirel. Commun. (2020)
13. Charan, G., Osman, T., Hredzak, A., Thawdar, N., Alkhateeb, A.: Vision-position multi-modal beam prediction using real millimeter wave datasets. In: IEEE Wireless Communications and Networking Conference (WCNC), pp. 2727 – 2731 (2022)
14. Demirhan, U., Alkhateeb, A.: Radar aided 6G beam prediction: deep learning algorithms and real-world demonstration. In: IEEE Wireless Communications and Networking Conference (WCNC), pp. 2655–2660 (2022)
15. Yang, R., Zhang, Z., Zhang, X., Li, C., Huang, Y., Yang, L.: Meta-learning for beam prediction in a dual-band communication system. IEEE Trans. Commun. **71**(1), 145–157 (2023)
16. Xie, T., Dai, L., Ng, D.W.K., Chae, C.B.: On the power leakage problem in millimeter-wave massive MIMO with lens antenna arrays. IEEE Trans. Signal Process. **67**(18), 4730–4744 (2019)
17. Jiang, T., Cheng, J.: Target recognition based on CNN with leakyReLU and PReLU activation functions. In: International Conference on Sensing, Diagnostics, Prognostics, and Control (SDPC), pp. 718–722 (2019)
18. He, K., Zhang, X., Ren, S., Sun, J.: Deep residual learning for image recognition. In: IEEE Conference on Computer Vision and Pattern Recognition (CVPR), pp. 770–778 (2016)
19. Simonyan, K., Zisserman, A.: Very deep convolutional networks for large-scale image recognition. In: International Conference on Learning Representations (2015)

Chapter 6
Conclusion and Outlook

Abstract This chapter first concludes the work of this monograph, i.e., AI for wireless physical layer, including intelligent channel estimation and interpolation, CSI feedback, precoding as well as spatial beam management techniques. Then, several future research directions are given, such as intelligent transceiver, MAC, RRC, core network, and network security.

Keywords AI · Wireless physical layer · Transceiver · Network

6.1 Conclusion

This monograph aims to elaborate that AI-based physical layer technology can be utilized to provide high Quality of Service (QoS) for 6G communications and future services. The AI-based channel estimation and interpolation techniques are first studied and analyzed in detail. Then, the intelligent transmission technologies under FDD and TDD are investigated in detail, respectively, i.e., CSI feedback and precoding are discussed, demonstrating the feasibility and efficiency of AI technology in these two scenarios. The spatial beam management techniques based on AI is finally summarized and compared in detail. It is concluded that AI4NET can be used to efficiently support information transmission in 6G networks.

6.2 Future Research Directions

The investigation of AI-based physical layer technology has stimulated strong research interest in both academia and industry alike. However, much more efforts are needed for practical applications of AI4NET in the near future. Based on our discussions in the preceding chapters, the following challenges and open issues lie ahead.

© The Author(s), under exclusive license to Springer Nature Switzerland AG 2025   85
L. Zhao et al., *AI for Wireless Physical Layer*, SpringerBriefs in Computer Science,
https://doi.org/10.1007/978-3-032-01367-5_6

Intelligent Receiver While AI technology has significantly enhanced various aspects of receiver, including channel estimation [1], signal equalization, demodulation [2], and decoding [3], critical challenges remain in receiver design. The development of integrated end-to-end AI-based receivers—capable of directly transforming received waveforms into transmitted bits or higher-level semantic information—requires further investigation to achieve optimal performance [4–6]. Moreover, research efforts should prioritize energy-efficient implementations, particularly lightweight AI architectures tailored for low-power devices, to enable practical deployment in resource-constrained environments like Internet of Things (IoT) networks and edge computing systems.

Intelligent Transmitter While AI technology has demonstrated remarkable advancements in optimizing user scheduling algorithms and transmit precoding systems, three critical research frontiers in intelligent transmitter design warrant further exploration: (1) context-aware waveform design, where reinforcement learning dynamically optimizes waveforms based on real-time channel states and user demands [7]; (2) joint transceiver optimization, leveraging end-to-end neural networks to codesign transmitter and receiver architectures for enhanced system synergy [8, 9]; and (3) green transmitters, employing AI-driven energy-efficient strategies such as adaptive power allocation and sleep-mode optimization to minimize energy consumption and environmental impact.

Intelligent Media Access Control (MAC) Technology Beyond intelligent physical layer advancements, intelligent MAC technology merits dedicated research attention across the following key dimensions: (1) Self-organizing MAC protocols employing RL for dynamic spectrum access in heterogeneous network environments (e.g., Unmanned Aerial Vehicle (UAV) swarms, IoT deployments) [10]; (2) Predictive resource allocation leveraging advanced time-series forecasting models (e.g., LSTM, transformers) to enable proactive resource scheduling; (3) Collision-free MAC architectures utilizing Graph Neural Networks (GNNs) for distributed collision avoidance in ultradense network deployments; (4) Cross-layer intelligent MAC optimization through tight integration of Physical Layer (PHY)-layer CSI with AI-optimized MAC-layer decision processes to achieve joint performance-efficiency enhancement.

Intelligent Radio Resource Control (RRC) Technology To advance AI-driven network resource efficiency [11], the following research directions warrant in-depth investigation: (1) Intelligent RRC state control could be developed to optimize idle-connected transition mechanisms in order to minimize signaling overhead and energy consumption; (2) Predictive mobility management could leverage spatiotemporal neural networks for enhanced anticipatory handover decision-making in multicell environments [12]; (3) Online reinforcement learning could be designed for service-aware RRC to dynamically adapt parameters (e.g., Discontinuous Reception (DRX) cycles, timers) across diverse QoS profiles; (4) Slice-optimized RRC mechanisms can design differentiated state machines for eMBB / URLLC / mMTC slices via AI-optimized parameter adaptation [13, 14].

Intelligent Core Networks Key research directions for intelligent core networks primarily encompass: (1) Self-healing network mechanisms for automated fault prediction and recovery [15]; (2) Intent-driven networking architectures translating business objectives into network configurations [16]; (3) AI-native architectures with embedded ML/AI as foundational components in 6G core networks; (4) Quantum ML applications for ultralow latency routing and resource allocation [17]. These innovations could collectively enhance network resilience, adaptability, and operational efficiency.

Network Security Future research of network security should concentrate on five interconnected pillars: (1) Distributed privacy preservation through cross-device federated learning with gradient encryption for edge network collaboration [18]; (2) PHY-layer adversarial defense systems employing certified robustness against model poisoning and gradient inversion attacks; (3) Secure MAC protocol codesign integrating federated learning with homomorphic encryption for traffic-obscured coordination; (4) RRC-layer security enhancement via graph neural network-based detectors combating false base station attacks and signaling anomalies [19]; (5) Blockchain-empowered trust architectures combining zero-knowledge proofs and smart contracts for verifiable AI model governance. These synergistic advancements establish a multilayered defense framework that could enable resilient, privacy-guaranteed 6G intelligent networks [20].

References

1. Zimaglia, E., Riviello, D.G., Garello, R., Fantini, R.: A deep learning-based approach to 5G-new radio channel estimation. In: 2021 Joint European Conference on Networks and Communications & 6G Summit (EuCNC/6G Summit) (2021), pp. 78–83
2. He, D., Wang, Z.: Deep learning-assisted demodulation for terahertz communications under hybrid distortions. IEEE Commun. Lett. **26**(2), 325–329 (2022)
3. Kee, H.L.M., Ahmad, N., Izhar, M.A.M., Anwar, K., Ng, S.X.: A review on machine learning for channel coding. IEEE Access **12**, 89002–89025 (2024)
4. Pihlajasalo, J., Korpi, D., Honkala, M., Huttunen, J.M., Riihonen, T., Talvitie, J., Brihuega, A., Uusitalo, M.A. Valkama, M.: Deep learning OFDM receivers for improved power efficiency and coverage. IEEE Trans. Wirel. Commun. **22**(8), 5518–5535 (2023)
5. Honkala, M., Korpi, D., Huttunen, J.M.J.: Deeprx: fully convolutional deep learning receiver. IEEE Trans. Wirel. Commun. **20**(6), 3925–3940 (2021)
6. Cammerer, S., Aoudia, F.A., Hoydis, J., Oeldemann, A., Roessler, A., Mayer, T., Keller, A.: A neural receiver for 5G NR multi-user MIMO. In: 2023 IEEE Globecom Workshops (GC Wkshps) (2023), pp. 329–334
7. Koivunen, V., Keskin, M.F., Wymeersch, H., Valkama, M., González-Prelcic, N.: Multicarrier ISAC: advances in waveform design, signal processing, and learning under nonidealities. IEEE Signal Process. Mag. **41**(5), 17–30 (2024)
8. Xue, S., Ma, Y., Yi, N.: End-to-end learning for uplink MU-SIMO joint transmitter and non-coherent receiver design in fading channels. IEEE Trans. Wirel. Commun. **20**(9), 5531–5542 (2021)
9. Xue, S., Ma, Y., Yi, N., Tafazolli, R.: Unsupervised deep learning for MU-SIMO joint transmitter and noncoherent receiver design. IEEE Wirel. Commun. Lett. **8**(1), 177–180 (2019)

10. He, L., Hu, F., Chu, Z., Zhao, J., Sagduyu, Y., Thawdar, N., Kumar, S.: Intelligent terahertz medium access control (MAC) for highly dynamic airborne networks. IEEE Trans. Aerospace Electron. Syst. **59**(3), 2494–2512 (2023)
11. Liang, L., Ye, H., Yu, G., Li, G.Y.: Deep-learning-based wireless resource allocation with application to vehicular networks. Proc. IEEE **108**(2), 341–356 (2020)
12. Paropkari, R.A., Thantharate, A., Beard, C.: Deep-mobility: a deep learning approach for an efficient and reliable 5G handover. In: 2022 International Conference on Wireless Communications Signal Processing and Networking (WiSPNET) (2022), pp. 244–250
13. Hu, T., Liao, Q., Liu, Q., Carle, G.: Information bottleneck-based domain adaptation for hybrid deep learning in scalable network slicing. IEEE Trans. Mach. Learn. Commun. Netw. **2**, 1642–1660 (2024)
14. Wu, W., Zhou, C., Li, M., Wu, H., Zhou, H., Zhang, N., Shen, X.S., Zhuang, W.: AI-native network slicing for 6G networks. IEEE Wirel. Commun. **29**(1), 96–103 (2022)
15. Arulappan, A., Mahanti, A., Passi, K., Srinivasan, T., Naha, R., Raja, G.: DQN approach for adaptive self-healing of VNFs in cloud-native network. IEEE Access **12**, 34489–34504 (2024)
16. Li, T., Yang, C., Song, Y., Cai, L., Zheng, R., Liu, X., Ji, Z., Liu, S.: Architecting autonomous network management and control via intent-driven decoupled network. IEEE Network **38**(6), 361–369 (2024)
17. Duong, T.Q., Ansere, J.A., Narottama, B., Sharma, V., Dobre, O.A., Shin, H.: Quantum-inspired machine learning for 6G: Fundamentals, security, resource allocations, challenges, and future research directions. IEEE Open J. Veh. Technol. **3**, 375–387 (2022)
18. Sandeepa, C., Zeydan, E., Samarasinghe, T., Liyanage, M.: Federated learning for 6G networks: navigating privacy benefits and challenges. IEEE Open J. Commun. Soc. **6**, 90–129 (2025)
19. Kuadey, N.A.E., Maale, G.T., Kwantwi, T., Sun, G., Liu, G.: Deepsecure: detection of distributed denial of service attacks on 5G network slicing—deep learning approach. IEEE Wirel. Commun. Lett. **11**(3), 488–492 (2022)
20. Rahman, M.A., Hossain, M.S.: A deep learning assisted software defined security architecture for 6G wireless networks: IIoT perspective. IEEE Wirel. Commun. **29**(2), 52–59 (2022)

The manufacturer's authorised representative in the EU is Springer
Nature Customer Service Centre GmbH, Europaplatz 3, 69115 Heidelberg,
Germany. If you have any concerns regarding our products, please
contact ProductSafety@springernature.com

Printed and bound by CPI Group (UK) Ltd, Croydon, CR0 4YY

28/04/2026

02098544-0003